Series on the Foundations of Natural Science

Nano-Engineering in Science and Technology

An Introduction to the World of Nano-Design

Michael Rieth
AIFT, Karlsruhe, Germany

Published by

World Scientific Publishing Co. Pte. Ltd.
P O Box 128, Farrer Road, Singapore 912805
USA office: Suite 1B, 1060 Main Street, River Edge, NJ 07661
UK office: 57 Shelton Street, Covent Garden, London WC2H 9HE

British Library Cataloguing-in-Publication Data
A catalogue record for this book is available from the British Library.

Series on the Foundations of Natural Science and Technology – Vol. 6
NANO-ENGINEERING IN SCIENCE AND TECHNOLOGY
An Introduction to the World of Nano-Design

ISBN 981-238-073-6
ISBN 981-238-074-4 (pbk)

Printed by FuIsland Offset Printing (S) Pte Ltd, Singapore

Nano-Engineering in Science and Technology

An Introduction to the World of Nano-Design

Series on the Foundations of Natural Science and Technology

Series Editors: C. Politis (*University of Patras, Greece*)
W. Schommers (*Forschungszentrum Karlsruhe, Germany*)

Vol. 1: Space and Time, Matter and Mind: The Relationship between Reality and Space-Time
by W. Schommers

Vol. 2: Symbols, Pictures and Quantum Reality: On the Theoretical Foundations of the Physical Universe
by W. Schommers

Vol. 3: The Visible and the Invisible: Matter and Mind in Physics
by W. Schommers

Vol. 4: What is Life? Scientific Approaches and Philosophical Positions
by H.-P. Dürr, F.-A. Popp and W. Schommers

Vol. 5: Grasping Reality: An Interpretation-Realistic Epistemology
by H. Lenk

Preface

The idea of building unimaginable small things at the atomic level is nothing new. Already in 1959, *R. Feynman*, the 1965 *Nobel* prize winner in physics, described during his famous dinner talk, "There's plenty of room at the bottom!" how it might be possible to print the whole 24 volumes of the *Encyclopedia Brittanica* on the head of a stick pin. He even speculated on how to store information at atomic levels or how to build molecular-sized machines:

"*I am not afraid to consider the final question as to whether, ultimately in the great future we can arrange atoms the way we want; the very atoms, all the way down! ⋯ The principles of physics, as far as I can see, do not speak against the possibility of maneuvering things atom by atom. It is not an attempt to violate any laws ⋯ but in practice, it has not been done because we are too big ⋯ The problems of chemistry and biology can be greatly helped if our ability to see what we are doing, and to do things on an atomic level, is ultimately developed — a development which I think cannot be avoided*". [Feynman, 1960]

Now, some decades later, new laboratory microscopes can not only visualize but manipulate individual atoms. With this recently developed ability to measure, manipulate and organize matter on the atomic scale, a revolution seems to take place in science and technology. And unfortunately, wherever structures smaller than one micrometer are considered the term *nanotechnology* comes into play. But nanotechnology comprises more than just another step toward miniaturization!

While nanotechnology may be simply defined as technology based on the manipulation of individual atoms and molecules to build structures to complex atomic specifications [Policy Research Project, 1989], one has to consider further that at the nanometer scale qualitatively new effects, properties and processes emerge which are dominated by quantum mechanics, material confinement in small structures, interfacial volume fraction, and other phenomena. In addition, many current theories of matter at the micrometer scale have critical lengths of nanometer dimensions and therefore, these theories are not adequate to describe the new phenomena at the nanometer scale.

Nevertheless, the concept of nanotechnology goes much further. It is an anticipated manufacturing technology giving thorough, inexpensive control of the structure of matter where other terms, such as molecular manufacturing, nano-engineering, etc. are also often applied. In other words, the central thesis of nanotechnology is that almost any chemically stable structure that can be specified can in fact be built. Researchers hope to design and program nano-machines that build large-scale objects atom by atom. With enough of these *assemblers* to do the work, along with *replicators* to build copies of themselves, we could manufacture objects of any size and in any quantity using common materials like dirt, sand, and water [Drexler, 1981; Drexler *et. al*, 1991; Regis, 1995; Merkle, 2001]. Computers 1000 times faster and cheaper than current models; biological nano-robots that fix cancerous cells; towers, bridges, and roads made of unbreakable diamond strands; or buildings that can repair themselves or change shape on command might be futuristic but likely implications of nanotechnology.

Today, while nanotechnology is still in its infancy and while only rudimentary nanostructures can be created with some control, this seems like science fiction. But respected scientists agree that it is possible, and more and more of the pieces needed to do it are falling into place. Nanotechnology has captured the imaginations of scientists, engineers and economists not only because of the explosion of discoveries at the nanometer scale, but also because of the potential societal implications. A White House letter (from the Office of Science and Technology Policy and Office of Management and Budget) sent in the fall of 2000 to all Federal agencies has placed nanotechnology at the top of the list of emerging fields of research and development in the United States. The *National Nanotechnology Initiative* was approved by Congress in November 2000, providing a total of $422 million spread over six departments and agencies [NNI; Roco, Sims, 2001]. And this certainly doesn't seem like science fiction!

Now, let us discuss nanotechnology from the educational point of view. What might be the most important scientific branch with respect to the development of nanotechnological applications?

To apply nanotechnology, researchers have to understand biology, chemistry, physics, engineering, computer science, and a lot of other special topics, such as protein engineering or surface physics. But the complexity of modern science forces scientists to specialize and the exchange of information between different disciplines is unfortunately not very common. So the breadth is one of the reasons why nanotechnology proves so difficult to develop.

But even today, one tendency is clearly visible: nanotechnology makes *design* the most important part of any development process. If nanotechnology comes true, the traditional production costs would drop to almost nothing, while the amount of design work would increase enormously due to its complexity. Further, the field of engineering design will become much more complex. Someone has to design these atomic-sized assemblers and replicators as well as nano-materials and others. And if we can build anything in any quantity, the practical question of "What can we build?" becomes a philosophical one: "What do we choose to build?". And this in turn is a design question. Answering it and planning for the widespread change each nano design could bring makes design planning incredibly important [Milanski, 2000].

As a conclusion, we may summarize: design will change radically under nanotechnology and for nano-engineers or nano-designers, respectively, a broad knowledge will become even more important in the future.

As long as we are still far away from the realization of complex nanotechnological applications, nano-engineering and nano-design almost exclusively take place on computers. Computational nano-engineering is an important field of research aimed at the development of nanometer scale modeling and simulation methods to enable and accelerate the design and construction of realistic nanometer scale devices and systems. Comparable to micro-fabrication which has led to the microelectronics revolution in the 20th century, nano-engineering and design will be a key to the nanotechnology revolution in the 21st century.

Therefore, the intention of this monograph is to give an introduction into the procedures, techniques, problems and difficulties arising with computational nano-engineering and design.

For the sake of simplicity, the focus is laid on the *Molecular Dynamics method* which is well suited to explain the topic with just a basic knowledge of physics. Of course, at some points we have to go further into detail, i.e. quantum mechanics or statistical mechanics knowledge is needed. But such subsections may be skipped without loosing the picture.

I am particularly grateful to W. Schommers (Editor) for his encouragement, assistance and advice. I also thank F. Schmitz for his support in all matters of high performance computation. Further, I am grateful to E. Materna–Morris for preparing the SEM pictures. A special thanks goes to Natascha for her careful reading and checking of the manuscript and to Rebecca for her moral support. I am indebted to C. Politis and numerous other persons for many interesting discussions on the topic. Last but not least, I would like to thank S. Patt (Editor) and the entire team from World Scientific for the close and professional collaboration during the publication of this book.

Michael Rieth
Karlsruhe, 2002

Contents

Chapter 1

Introduction

Today, nanotechnology is still at the beginning, and only rudimentary nanostructures can be created with some control. The science of atoms and simple molecules, on one end, and the science of matter from microstructures to larger scales, on the other, are generally established. The remaining size-related challenge is at the nanometer scale — roughly between 1 and 100 molecular diameters — where the fundamental properties of materials are determined and can be engineered. A revolution has been occurring in science and technology, based on the developed ability to measure, manipulate and organize matter on this scale. Recently discovered organized structures of matter (such as carbon nano-tubes, molecular motors, DNA-based assemblies, quantum dots, and molecular switches) and new phenomena (such as giant magnetoresistance, coulomb blockade, and those caused by size confinement) are scientific breakthroughs that merely hint at possible future developments [Roco, Sims, 2001].

More and more, small structures with dimensions in the nanometer regime play an important role within molecular biology, chemistry, materials science and solid-state physics.

Of particular interest in biology there is, for example, the replication of proteins, the functionality of special molecular mechanisms like haemoglobin or even such seemingly simple structures like the flagella of certain bacteria. Chemistry, on the other hand, deals with the synthesis — and therefore also with an improvement — of these structures with which nature solves so many problems. For example, the design of catalysts is a considerable commercial factor within the chemical industry. Specific modifications of properties of well-known materials using small particles and the development of fabrication processes of nano-particles are topics of modern material sciences. Self-cleaning surfaces as well as pigments are typical examples for applications of nanostructures where, interestingly, the latter already led to some success within the cosmetic industry [Siegel, 1997].

But nanotechnology comprises more than just producing small structures — the concept goes much further. Nanotechnology is an anticipated manufacturing technology giving thorough, inexpensive control of the structure of matter where other terms, such as *molecular manufacturing*, *nano-engineering*, etc. are also often applied. Researchers hope to design and program *nano-machines* that build large-scale objects atom by atom. With such self-replicating *assemblers* objects of any size and in any quantity could be manufactured using common materials like dirt, sand, and water. Computers 1000 times faster and cheaper than current devices; biological nano-robots that fix cancerous cells; towers, bridges, and roads made of unbreakable diamond strands; or buildings that can repair themselves or change shape on command might be future but likely implications of nanotechnology.

What makes nanostructures different? They show significantly different properties compared to the bulk material. As is known from quantum mechanics the electronic states of nano-particles are considerably changed compared to the bulk. This is due to quantization effects caused by the spatial restriction. The electronic structure, on the other hand, is responsible for all those material properties like electronic conductivity, optical absorption, chemical reactivity or even the mechanical properties. Therefore, these nanostructures appear as particles with new material properties [Jena *et. al*, 1987].

The investigation of nanostructures is a highly topical field of solid state physics and materials research. New, sophisticated characterization methods have been successfully developed during the last twenty years like the scanning tunnelling microscope (STM), for example, which has been established as standard instrument for scanning nanostructures on surfaces or the transmission electron microscope (TEM) combined with theoretical modeling for visualization of periodic structures. Even scattering methods (ions, electrons, X-rays, neutrons) have been improved to an extend which is hard to beat. Finally, spectroscopic information with high resolution has become available through the use of synchrotron radiation sources of the third generation [FZJ, 1998].

Beside all these experimental characterization techniques, which are applicable to existing structures only and which are most often time and cost intensive, computational methods for many-particle systems have made their entrance into all branches of science for which the term nanotechnology has been established. Computer experiment, computer chemistry, molecular design, nano-machinery, nano-manufacturing and nano-computation are just a few subjects which have come up in connection with numerical calculations in the field of nanotechnology [Alig *et. al*, 2000].

Here, one tendency is clearly recognizable: nanotechnology makes *design* the most important part of any development process. With nanotechnology the amount of design work increases enormously due to its complexity. Planning for the widespread change, each nano-design could make design planning incredibly important [Milanski, 2000]. To summarize, design will change radically under nanotechnology and for nano-engineers or nano-designers, respectively, a broad knowledge will become even more important in the future.

Trying to categorize the numerical solution techniques for many-particle systems basically leads to four different topics: quantum theoretical calculations (*ab initio*), molecular mechanics, Monte Carlo, and molecular dynamics methods.

While the solution of *Schrödinger's* equation for many-particle systems is inherently impossible — the calculation time increases exponentially with the particle number — quantum theoretical calculation methods focus on approximation and separation approaches to simplify the calculation scheme. Some of the most common *ab initio* methods are self-consistent field methods, the linear combination of atomic orbitals or the density functional method [Sauer, 2000].

In contrast to *ab initio* methods, molecular mechanics and molecular dynamics are based on classical mechanics. The particles are treated as mass points interacting through force fields which in turn are derived from interacting potentials. The goal of molecular mechanics (as well as of *ab initio* calculations) is to find stable configurations for a set of particles, that is, to determine saddle points (local minima) on the potential energy surface. While quantum mechanical calculations lack an *a priori* concept of chemical bonds, molecular mechanic methods use the approach, known from traditional organic chemistry, where molecules are characterized by ball-and-stick models in which each ball represents an atom and each stick represents a bond. Depending on the kind of bond, appropriate interaction potentials have to be chosen and, therefore, energy functions and parameters have to be tailored to specific local arrangements of atoms. In this way molecular mechanics programs treat the potential energy as a sum of terms accounting chiefly for bond stretching, bending, torsion and for *van-der Waals*, overlap and electrostatic interactions among non-bonded atoms. Molecular mechanics systems have, however, been successfully applied to just a narrow range of molecular structures in configurations not too far from equilibrium [Drexler, 1992].

Similar considerations are valid for molecular dynamics calculations. But in contrast to Monte Carlo methods where new particle configurations are created randomly step by step, molecular dynamics works through the solution of *Newton's equations of motion*. Therefore, the evolution of a many-particle system can be calculated in certain time steps where the total information (particle positions, velocities, kinetic and potential energies, etc.) of the system is available for each time step. All further properties — like for example the temperature — can be determined without any additional parameters.

This is not the case for Monte Carlo methods. Here one generally samples system configurations according to a given statistical ensemble, characterized by *Boltzmann* distributions which include the temperature as external parameter and, therefore, such calculations are only applicable for configurations near the equilibrium. Additional problems arise in the attempt to assign time steps to the different configurations [Ciccotti *et. al*, 1986]. *Ab initio* calculations also lack the subject temperature by nature, because there are no dynamic considerations involved.

Beside this, each of the four calculation techniques has its advantages as well as limitations. When performing computational methods the results should basically mirror reality as closely as possible. *Ab initio* calculations work without additional *a priori* input like interaction potentials and — depending on the degree of simplification used in the particular method — the results include explicitly several different quantum effects. On the other hand, the computational effort is enormous, i.e. usually the systems are restricted to less than a few hundred atoms. Nevertheless, these methods have revolutionized chemistry with the computer aided design of molecules among many other applications.

While both, molecular mechanics and molecular dynamics methods, are based on classical many-particle physics, there are no explicit results from quantum effects available. Furthermore, these methods need a detailed knowledge of the particle interactions before the numerical calculation can be started, that is, specific models have to be established differing from case to case and depending on the study. Here, quantum mechanics comes into play implicitly with the use of interaction potentials, gained, for example, from *ab initio* calculations. Most often additional fits of such potentials to experimental data are necessary to obtain realistic results.

However, the precision and validity of the interaction potentials within molecular mechanics and molecular dynamics calculations restrict the field of application of these methods. On the other hand, both methods are able to handle large systems with about 10^5 to 10^7 atoms depending on the study.

Most modern commercial molecular mechanics programs use libraries of phenomenological potentials to describe all the different types of interactions occurring in the field of organic chemistry. With these it is possible to study minimum-energy configurations, stiffness, bearing and other properties of nanostructures (molecules), which are built largely of carbon atoms joint by strong, directional, covalent bonds (single, double, triple, hybrid) which in turn are often augmented with one or more different elements. Due to the simplified description of the atomic interactions — aside from the small inaccuracies found in all structures — standard molecular mechanic programs cannot realistically describe certain structures. For example, they can model many stable structures, even when strained, but they cannot describe chemical transformations or systems which are close to the transformation point. Therefore, computational results must be examined closely for such invalid conditions. However, studies for broad classes of organic structures including large biomolecules as well as polymers are possible with a computation cost favor by a factor of more than 10^3 compared to *ab initio* methods [Drexler, 1992].

Since molecular dynamics methods are more sensitive to inappropriate forces — with respect to the validity of the results — it is even more important to concentrate on the use of properly determined interaction potentials. It is absolutely necessary to consider the range of validity, the applicability as well as the accurateness of the underlying interaction potentials whenever molecular dynamics methods are applied [Gehlen *et. al*, 1972].

While most works use either many-body forces (for the description of covalent bonds) or phenomenological inter-atomic potentials, in this book, in contrast, we focus mainly on mono-atomic nanosystems for which reliable, precise interaction forces are available within a wide range of applicability. To be more specific, we restrict ourselves — as far as possible — to studies using exclusively one of the two materials: a noble gas (krypton) and a simple metal (aluminium).

At first glance, however, this seems not to promise spectacular results, but — as will be shown later on — even seemingly simple nanostructures most often do not behave like they are assumed to do. While this is typical for the whole field of nanotechnology, the focus within the present monograph is laid on such basic "nano-effects" which can only be detected by the use of realistic descriptions of the atomic interactions. On the other hand, more complicated scenarios like nano-machines with metallic parts will be outlined, too.

Finally, it should be emphasized that working within computational nano-physics by means of molecular dynamics implies a combination of several scientific fields like atomic interaction potential theory (which in turn is a combination of several different branches of theoretical and experimental physics), computer science and statistical mechanics.

Therefore, we start with a brief introduction into atomic potentials for noble gases and simple metals and then continue with an excursus through the field of molecular dynamics and nano-design which is followed by a review of several characterization functions known from statistical mechanics. Finally, these introducing chapters are succeeded by presentations and discussions of different application examples and studies which provide an insight into the world of computational nano-engineering.

Chapter 2

Interatomic Potentials

2.1 Quantum Mechanical Treatment of the Many-Particle Problem

The quantum mechanical modeling of a system with N particles of masses m_i leads to the *Hamiltonian*

$$\hat{H} = \sum_{i=1}^{N} \left[-\frac{\hbar^2}{2m_i} \nabla_i^2 + V_i(\mathbf{r}_i) \right] + \sum_{\substack{i,k=1 \\ i \neq k}}^{N} V_{ik}(\mathbf{r}_i, \mathbf{r}_k) \,. \tag{2.1}$$

Here, $V_i(\mathbf{r}_i)$ is an externally given potential in which the ith particle is located and $V_{ik}(\mathbf{r}_i, \mathbf{r}_k)$ denotes the interaction potential between the two particles i and k. To analyze or to describe its characteristics, one has to solve the corresponding many-particle *Schrödinger* equation

$$\hat{H}\Psi = E\Psi \,, \tag{2.2}$$

where E is the total energy. The wave function Ψ depends on the $3N$ co-ordinates (*configuration space*) of all particles:

$$\Psi = \Psi(x_1, y_1, z_1, \cdots, x_i, y_i, z_i, \cdots, x_N, y_N, z_N) \,. \tag{2.3}$$

If we consider nanosystems, most often external potentials are not present and the particles involved are atoms which in turn have to be divided into nuclei (N) and electrons (e). In this case, the interaction potential of Eq. 2.1 is given by the *Coulomb* potential

$$V_{ik}(\mathbf{r}_i, \mathbf{r}_k) = \frac{Z_i Z_k \; e^2}{|\mathbf{r}_k - \mathbf{r}_i|} \,, \tag{2.4}$$

where Z is the electron charge number including the sign of the charge.

With a closer look at this many-particle problem, it becomes clear that an exact quantum mechanical solution can probably never be achieved. Here is an example: a relatively small nano-cluster of only 100 argon atoms consists of 100 nuclei and 1800 electrons, which is a total of 1900 particles. In this case, the *configuration space* consists of 5700 dimensions. The key point for numerical solutions of the *Schrödinger* equation is the spatial integration. With the assumption that a division of each dimension into 100 steps is sufficient for an accurate calculation, we would have to compute the summation of 10^{11400} volume elements. It is needless to mention that this is not possible without further intensive simplifications and approximations.

Therefore, quantum theoretical calculation methods (*ab initio* or *first principle*, respectively) mainly focus on approaches that reduce the dimensions of the configuration space. One of the most common approaches is valid under the condition that the electrons have a much higher kinetic energy than the nuclei. While that is certainly true for most nanotechnological considerations the procedure, known as *Born–Oppenheimer* [Born, Oppenheimer, 1927] or *adiabatic approximation* [Messiah, 1990], consists of separating the electron and nuclear motions (wave functions) and treating each independently. Then the wave function (Eq. 2.3) can be written in a slightly more manageable form (with m nuclei and n electrons):

$$\Psi = \phi_N \, \psi_e = \phi_N(x_{N1}, y_{N1}, z_{N1}, \cdots, x_{Nm}, y_{Nm}, z_{Nm}) \times \psi_e(x_{e1}, y_{e1}, z_{e1}, \cdots, x_{en}, y_{en}, z_{en}) \,, \tag{2.5}$$

where the electron wave function still depends on the nuclear positions.

A far more effective reduction of the problem can be achieved if all the electrons are bound to a central field as is the case within a single atom. Here one of the most important *ab initio* methods — the *self-consistent field method* [Messiah, 1990; Greiner, 1993; Landau, Lifshitz, 1959] — goes one step further. The idea of this method is to regard each electron of an atom as being in motion in the combined field due to the nucleus together with all the other electrons (*self-consistent field*). In this way, the central *Coulomb* field of the nucleus appears as pseudo external potential within the *Hamiltonian* and the highly dimensional combined wave function of the electrons ψ_e is separable into the according single wave functions of just three spatial dimensions for each electron:

$$\psi_e = \psi_1(x_1, y_1, z_1)\, \psi_2(x_2, y_2, z_2) \cdots \psi_n(x_n, y_n, z_n)\,. \tag{2.6}$$

The method is named after *Hartree* [Hartree, 1955] and works by iterative calculation of the single electron *Schrödinger* equations and of the medium field due to all electrons until self-consistency is reached. But despite its simplicity the method has some disadvantages. The wave function (Eq. 2.6) is not anti-symmetric, i.e. one has to take care of impossible configurations, e.g. by putting each electron into another state to fulfill *Pauli's principle*. Another problem is the necessity of ortho-normalizing the wave functions during the iteration loops.

With the *Hartree–Fock* method [Fock, 1930] proper anti-symmetric and permanently ortho-normal wave functions have been introduced into the *Hartree* scheme by arranging the single electron wave functions — including electron spin s — in the way of *Slater's determinant*:

$$\psi_e = \frac{1}{\sqrt{n!}} \begin{vmatrix} \psi_1(\mathbf{r}_1, s_1) & \psi_2(\mathbf{r}_1, s_1) & \cdots & \psi_n(\mathbf{r}_1, s_1) \\ \psi_1(\mathbf{r}_2, s_2) & \psi_2(\mathbf{r}_2, s_2) & \cdots & \psi_n(\mathbf{r}_2, s_2) \\ \vdots & \vdots & & \vdots \\ \psi_1(\mathbf{r}_n, s_n) & \psi_2(\mathbf{r}_n, s_n) & \cdots & \psi_n(\mathbf{r}_n, s_n) \end{vmatrix}\,. \tag{2.7}$$

The many-particle *Schrödinger* equation (Eq. 2.2) then becomes a system of non-local *Schrödinger* equations for single electrons. Beside the spin treatment, the *Hartree–Fock* method implicitly describes the exchange effect too. But other effects, like the electron correlations, are not included.

Another method uses the fact that the total energy of an atom — including the electron correlations — can be derived from the electron density. Mathematically, the spatial distribution of the electrons is far easier to handle compared to the wave function. In this way, the energy can be described as a *density functional*, which is the name of the method [Kohn, Sham, 1965].

But despite the significant reduction of the high dimensionality of the configuration space by these *ab initio* methods there are still lots of difficulties which have to be handled by further adaptations, simplifications, and approximations. Going further into detail would exceed the frame of this monograph. But after this brief explanation it should be clear that the quantum mechanical formulation of the many-particle problem is relatively simple, while its solution implies an enormous effort — even by restricting on approximative results.

2.2 Potential Energy Surface

Let us come back to the example with the nano-cluster of 100 argon atoms. As it is well known, there is a mutual (attracting) interaction of noble gas atoms due to the polarization of the "electron clouds" around the nuclei. What is the principle of a quantum mechanical calculation, if we are interested in the stable configurations of these 100 argon atoms?

In this case, the *Hamiltonian* for the $m = 100$ nuclei and $n = 1800$ electrons is

$$\begin{aligned}\hat{H} = \hat{T}_N + \hat{T}_e + V_{NN}(\mathbf{R}_1, \cdots, \mathbf{R}_m) + V_{ee}(\mathbf{r}_1, \cdots, \mathbf{r}_n) \\ + V_{eN}(\mathbf{r}_1, \cdots, \mathbf{r}_n, \mathbf{R}_1, \cdots, \mathbf{R}_m)\,.\end{aligned} \tag{2.8}$$

Here $\hat{T}_N$ is the kinetic energy operator for the nuclei

$$\hat{T}_N = -\sum_{i=1}^{m} \frac{\hbar^2}{2m_N} \frac{\partial^2}{\partial \mathbf{R}_i^2} \tag{2.9}$$

and $\hat{T}_e$ is the kinetic energy operator for the electrons

$$\hat{T}_e = -\sum_{j=1}^{n} \frac{\hbar^2}{2m_e} \frac{\partial^2}{\partial \mathbf{r}_j^2}\,. \tag{2.10}$$

According to the *Coulomb* interaction (Eq. 2.4) V_{eN} represents the attractive electron-nucleus, V_{ee} and V_{NN} the repulsive electron–electron and nucleus–nucleus interaction potential, respectively.

We now use the *Born–Oppenheimer approximation* (Eq. 2.5) to separate the wave function into a part ϕ for the nuclei which we assume to be "frozen" and a part ψ for the electrons:

$$\Psi(\mathbf{R}_1, \cdots, \mathbf{R}_m, \mathbf{r}_1, \cdots, \mathbf{r}_n) = \psi(\mathbf{R}_1, \cdots, \mathbf{R}_m, \mathbf{r}_1, \cdots, \mathbf{r}_n)\, \phi(\mathbf{R}_1, \cdots, \mathbf{R}_m)\,. \tag{2.11}$$

Therefore, the nuclear positions $\vec{\mathbf{R}} = [\mathbf{R}_1, \cdots, \mathbf{R}_m]$ within the electron wave function ψ appear as parameters only. Further, by neglecting the kinetic energy of the nuclei (*Born–Oppenheimer approximation*) we can easily write down the *Schrödinger* equation for the electrons:

$$\begin{aligned}\left[\hat{T}_e + V_{ee}(\mathbf{r}_1, \cdots, \mathbf{r}_n) + V_{eN}(\mathbf{r}_1, \cdots, \mathbf{r}_n, \vec{\mathbf{R}})\right] \psi(\mathbf{r}_1, \cdots, \mathbf{r}_n, \vec{\mathbf{R}}) \\ = \left[E_e(\vec{\mathbf{R}}) - V_{NN}(\vec{\mathbf{R}})\right] \psi(\mathbf{r}_1, \cdots, \mathbf{r}_n, \vec{\mathbf{R}})\,.\end{aligned} \tag{2.12}$$

For a given (fixed) nuclear configuration, Eq. 2.12 can be approximately solved, for example, by the use of the *self-consistent field* or the *density functional methods*, which in this general case have to be expanded further to handle the "multi-central field" configuration (in the descriptions of the preceding section we have considered the central field of just one nucleus). In this way, in principle, it is possible to gain the energy E_e for all possible configurations of the nuclei (here E_e is the electron energy plus *Coulomb* potential V_{NN} due to the nucleus–nucleus interaction).

For a better understanding of this result, Eqs. 2.8, 2.11 and 2.12 may now be substituted into the *Schrödinger* equation of the complete system. Performing the derivations and neglecting the mixed wave function terms — which would give rise to electron transitions between states (i.e. electron–phonon interaction) — leads by first approximation to the following equation:

$$[\hat{T}_N + E_e(\vec{\mathbf{R}})]\phi(\vec{\mathbf{R}}) = E\ \phi(\vec{\mathbf{R}})\ . \tag{2.13}$$

This is the *Schrödinger* equation for the nuclei where the energy of the electron states E_e acts as an effective potential for the nuclei. The interpretation with respect to our example is as follows: the 100 argon nuclei are moving in a "medium" caused by the 1800 electrons. It acts like a rubberband for the nuclei. Therefore, stable nuclear configurations can only appear at those points where the *potential energy surface* E_e shows minima (saddle points). Trying to find stable configurations for the argon cluster means localizing the minima of E_e with respect to the $m = 100$ co-ordinates of the atoms.

However, this is a brief and simple representation of the *ab initio* treatment of many-particle systems. There are so much different methods that even to mention them all would exceed this monograph. But basically there is one common principle as outlined above: (1) A calculation scheme for a certain point of the potential energy surface with more or less approximations, simplifications and adaptations according to the underlying study, and (2) an algorithm for localizing the minima. Due to their important role within the field of computational chemistry, most of the methods are available as commercial software packages [Clark, 1985] like, for example, TURBOMOLE [Ahlrichs, von Arnim, 1995]. The limit of modern *ab initio* methods combined with today's computer technology varies in the range of several hundred atoms strongly depending on the case of application. An example for aluminium clusters is given in [Ahlrichs, Elliott, 1999].

2.3 Pair Potential Approximation

As has just been shown for the example of an argon cluster (100 atoms), the according quantum mechanical many-particle problem can be reduced — with the help of the *Born–Oppenheimer approximation* — from 5700 dimensions (configuration space) into two parts of 5400 (electron wave function) and 300 (potential energy surface) dimensions, respectively. The electron wave function within the electron *Schrödinger* equation can be handled by a further reduction to a set of 1800 single electron wave functions, each with 3 spatial co-ordinates (without spin). This reductions are possible with the approximations that are based on the self-consistent field or density functional methods.

However, there is still the potential energy surface with 300 dimensions which cannot be calculated as a whole. But by looking for stable configurations only, the problem is reduced to the localization of its saddle points (minima). The faster a method is in homing in on a minimum, the less calculations of configuration points of the potential surface are necessary, and the more effective the underlying method is working.

But still, for calculations of larger systems a further reduction — comparable to that of the electron wave function with help of the *Hartree–Fock* method — is absolutely necessary. How can this be achieved?

Under the assumption that the change of the electronic arrangement around each atom may be considered as negligibly small within the considered system conditions, an expansion of the potential (energy surface) for N atoms can be applied:

$$E_e(\vec{\mathbf{R}}) = U(\mathbf{R}_1, \cdots, \mathbf{R}_N) = \frac{1}{2} \sum_{\substack{i,j=1 \\ i \neq j}}^{N} u_{ij} + \frac{1}{6} \sum_{\substack{i,j,k=1 \\ i \neq j \neq k}}^{N} u_{ijk} + \cdots . \tag{2.14}$$

Here the terms on the right in Eq. 2.14 represent pair, triplet and many-body contributions of the interatomic interactions. For neutral atoms, it is well known that the long-ranged parts of these interactions can be understood in terms of the resulting weakly attractive time averages of fluctuating and induced dipoles (*van-der Waals* or dispersion forces), whereas at short range the potentials tend to be quite strongly repulsive as a consequence mainly of the *exclusion principle*.

If the electron orbitals of the atoms are not easily polarizable, then, compared with the pair terms, the triplet and higher terms diminish rapidly in significance. The next step of approximation is to neglect them entirely.

This is called the *pair potential approximation*:

$$U(\mathbf{R}_1, \cdots, \mathbf{R}_N) = \frac{1}{2} \sum_{\substack{i,j=1 \\ i \neq j}}^{N} u_{ij} = \frac{1}{2} \sum_{\substack{i,j=1 \\ i \neq j}}^{N} v(|\mathbf{R}_i - \mathbf{R}_j|) \,. \qquad (2.15)$$

Referring to our example (100 argon atoms), with Eq. 2.15, the problem with the remaining 300 dimensions of the potential energy surface has been reduced to a 9900-fold sum of values from one pair potential function with only one dimension, which is the distance of two atoms. This simplification expands the calculability of the many-particle problem with today's computer power up to millions of particles — at least under certain conditions. Following *Neil Armstrong*, one could say: "That's one simple approximative step··· but one giant leap for the calculability of many-particle systems or nanostructures, respectively". On the other hand, such a far-reaching, if not to say brute simplification has a strong influence on the applicability, as can be easily imagined.

2.4 Advantages and Limitations of the Pair Potential Approximation

With the use of the pair potential concept the field of basic quantum mechanics is left very often, because it is rather difficult or even impossible to derive appropriate potential functions on the basis of *ab initio* methods. That's why most pair potentials are derived in a phenomenological way including, of course, quantum mechanical effects. Sometimes, as is the case for *pseudo potentials*, some parts are based on quantum mechanical considerations, others are fitted to experimental data. While *ab initio* methods can generally be applied without any additional *a priori* knowledge, working with pair potentials always implies the consideration of the specific conditions of the underlying study. Here the most critical and, of course, time consuming part is the derivation of a suitable pair potential function by using all the available data. On the other hand, if an appropriate function is at hand, the pair potential approximation considerably expands the area of applications. The calculability of the potential energy surface for thousands or even millions of particles opens the field for *molecular mechanics* as well as *molecular dynamics*.

Since the present monograph deals almost entirely with pairwise interatomic potentials, the fundamental question in this connection is that of the validity of the assumption that an assembly of atoms, be it solid, liquid, or gas, may be described in terms of two-body forces acting on pairs of atoms. From the quantum mechanical point of view, it would be of importance whether the wave functions of the atoms in the assembly are greatly altered by the interaction from their isolated values, or whether the modification may be considered as the result of a small perturbation of the free-atom function. In the latter case, the pair potential approximation is valid, whereas the former poses problems in the calculation of interaction potential energies [Torrens, 1972].

In general, atoms whose electron orbitals are saturated are not greatly disturbed when they approach each other up to the point of interpenetration. Unsaturated units, on the other hand, have their electronic structure substantially altered when they are mutually approaching. This would at first glance seem to limit the two-body potential model to a very narrow range of substances such as ionic crystals and *van-der Waals* solids. However, various approximations exist for dealing with metals (see *pseudo potentials*) and to some extent with covalent materials, which yield pair potentials valid under certain conditions [Torrens, 1972].

As mentioned above, if we are assuming a certain analytical form of interatomic potential and wish to obtain parameter values from experiments it is important to bear in mind the extent to which the pairwise interaction is valid in describing the physical phenomenon involved. The simple example of the use of experimental elastic constants serves to illustrate this point.

For the normal (simple) metals with small ion cores the ion–electron–ion interaction is predominant, while for noble or transition metals there is a significant repulsive force due to closed shell overlap in the solid. Nevertheless, in both cases there is a contribution to the elastic constants due to the conduction electron gas, which is volume-dependent and may not be completely described in terms of purely two-body forces [Torrens, 1972].

On the other hand, there are cases, in which certain three-body (dipole–dipole–dipole) terms in Eq. 2.14 can be incorporated into effective two-body terms, which as a result may become volume and temperature dependent [Casanova, 1970; Schommers, 1977, 1980]. In this case, the resulting potentials are referred to as *effective pair potentials*.

In connection with studies of metallic nanostructured systems, two additional considerations come into play:

(1) At free metal surfaces the local background electron density is different from its bulk value, and because the pair potential between the metal

ions depends critically on the electron density, the pair potential at the surface is different from that in the bulk of the crystal [von Blanckenhagen, Schommers, 1987]. But for clusters of a few nanometers in size the structure and dynamics of the atoms are dominated by surface atoms, i.e. interactions at the surface play a significant role. Therefore, an accurate description of metallic nanosystems by means of pair potentials necessitates the consideration of the electron density within the surface region and has to be described further in temperature dependent terms.

(2) Another basic question has to be answered prior to the use of pair potentials in connection with small metallic clusters. When at all does metallic behavior occur? In other words, how large has the cluster to be to show properties comparable to those in the solid bulk? It seems that clusters with at least 50 atoms can be considered as an approximate limit for the occurrence of metallic behavior [Eberhardt, 1998], i.e. this is a fundamental change in the pair potentials, which has to be kept in mind.

In conclusion, we may summarize that the pair potential concept works best in connection with noble gas solids. This is the only group of materials for which the pair potential functions are well known and at the same time are valid for a broad range of application. In contrast to other substances, the potential functions of noble gas solids do not depend on temperature and they are the same at the surface as well as in the bulk of the crystal. Therefore, parameters can be fitted by bulk properties derived from experimental results and the resulting *phenomenological potential* functions can be applied to both, surface and bulk area, without further adaptations.

2.5 Phenomenological Potentials

The present state of theoretical knowledge of pair potentials is such that phenomenologically derived potential functions often present a more realistic view of atomic interactions than potentials derived exclusively and usually with much efforts from purely theoretical considerations, which are themselves approximative in nature. Phenomenological atomic interactions are in most cases based on a simple analytical expression which may or may not be justifiable from theory and which contains one or more parameters adjusted to experimental results. By strict definition almost all forms of interatomic pair potentials which exist at present have to be described as *effective phenomenological potentials* due to the approximations necessary to overcome the many-body problem involved in the interaction [Torrens, 1972].

2.5.1 *Buckingham Potentials*

The original version of the *Buckingham* potential [Buckingham, 1938] has the form

$$v(r) = A\, e^{-\alpha \frac{r}{r_m}} - \frac{\lambda}{r^6} - \frac{\lambda'}{r^8}\,, \tag{2.16}$$

where

$$\left.\begin{aligned} A &= \left[-\varepsilon + (1+\beta)\tfrac{\lambda}{r_m^6}\right] e^{\alpha} \\ \lambda &= \tfrac{\varepsilon\, \alpha\, r_m^6}{\alpha(1+\beta)-8\beta-6} \\ \lambda' &= \beta\, r_m^6 \lambda \end{aligned}\right\} \tag{2.17}$$

with ε as the depth of the energy minimum and r_m as the corresponding value of the distance r between two atoms. The steepness of the exponential is measured by α, while β is the ratio of the inverse 8th to 6th power contributions at $r = r_m$. There are four independent constant parameters, which may be adjusted to experimental data. This potential, however, has the rather unphysical property of being negatively infinite at zero separation. Therefore, a variation known as the *Buckingham–Corner* potential eliminated the unrealistic behavior at the origin by postulating a more complicated form [Torrens, 1972; Buckingham, 1961].

A further variation, where the inverse 8th power term has been dropped, has led to a simpler form and is known as the *modified Buckingham* or *exponential*-6 potential:

$$v(r) = \frac{\varepsilon}{1-\frac{6}{\alpha}}\left[\frac{6}{\alpha}\, e^{\alpha\left(1-\frac{r}{r_m}\right)} - \left(\frac{r}{r_m}\right)^{-6}\right]. \tag{2.18}$$

There are three independent parameters (ε, r_m and α) which have the same significance to those of Eqs. 2.16 and 2.17. In connection with this potential there is one point which has to be considered: spuriously there is a maximum, usually for a very small separation $r = r_{\max}$. Therefore the *modified Buckingham potential* should only be applied to such calculations, where the energies are sufficiently small enough to avoid the region $r \leq r_{\max}$ [Torrens, 1972].

Table 2.1 Modified Buckingham potential parameter set for different non-bonded materials. The units are 10^{-7} kg, 10^{-21} J, and 10^{-10} m.

Symbol	Type	Mass	ε	α	r_m
C	sp, sp^2	19.925	0.357	12.5	3.88
C	sp^3, cycloprp.	19.925	0.357	12.5	3.80
H	hydrocarbon	1.674	0.382	12.5	3.00
H	alcohol	1.674	0.292	12.5	2.40
O	C–O–[H,C]	26.565	0.406	12.5	3.48
O	carbonyl	26.565	0.536	12.5	3.48
N	sp^3	23.251	0.447	12.5	3.64
F	fluoride	31.545	0.634	12.5	3.30
Cl	chloride	58.064	1.950	12.5	4.06
Br	bromide	131.038	2.599	12.5	4.36
I	iodide	210.709	3.444	12.5	4.64
S	sulphide	53.087	1.641	12.5	4.22
Si	silane	46.454	1.137	12.5	4.50
P	phosphine	51.464	1.365	12.5	4.36
Ne	noble gas	33.51	0.525	14.5	3.147
Ar	noble gas	66.34	1.701	14.0	3.866
Kr	noble gas	139.16	2.185	12.3	4.056

This potential is often used to describe the attractive and repulsive forces experienced by pairs of uncharged, non-bonded atoms. Table 2.1 shows a set of parameters for different atoms (the table is extracted from [Drexler, 1992], which in turn is a short version of [Burkert, Allinger, 1982], the noble gas parameters are from [Hirschfelder *et. al*, 1954]).

Further, a simple interpolation for the interaction of atoms from different materials can be applied by using the mean values of the according parameters:

$$\varepsilon_{12} = \frac{\varepsilon_1 + \varepsilon_2}{2}, \quad r_{m12} = \frac{r_{m1} + r_{m2}}{2} . \tag{2.19}$$

2.5.2 *Morse Potentials*

Morse proposed an interatomic potential without power law dependence in order to calculate the energy levels of diatomic molecules [Morse, 1929], wherein the potential should satisfy the following conditions [Torrens, 1972]:

(1) $v(r) \to 0$ as $r \to \infty$

(2) $\upsilon(r)$ has a minimum for $r = r_m$ (r_m represents the intermolecular separation)
(3) $\upsilon(r) \to \infty$ as $r \to 0$
(4) $\upsilon(r)$ should have the same allowed energy levels as those given by

$$E_n = -\varepsilon + \hbar\omega_0 \left[\left(n + \frac{1}{2} \right) - x \left(n + \frac{1}{2} \right)^2 \right] , \tag{2.20}$$

which describes the spectroscopic data of molecules.

Morse chose the following form for his potential:

$$\upsilon(r) = \varepsilon \left[e^{-2\alpha(r-r_m)} - 2e^{-\alpha(r-r_m)} \right] , \tag{2.21}$$

for which a solution of the radial part of the *Schrödinger* equation yields energy levels of the type given by Eq. 2.20. However, the *Morse* potential is not exclusively used for molecular energy level applications. It has been further extensively used in the study of lattice dynamics, the defect structure in metals, the inert gases in metals, the equation of state, elastic properties of metals, and the interaction between gas atoms and crystal surfaces, among many others. Table 2.2 shows a parameter set for several metals [Torrens, 1972; Girifalco, Weizer, 1959]. Since the minimum of these potentials are too deep for some considerations an additional row with a corrected value for ε according to the melting temperature ($\varepsilon_c = T_m\ k_b$) has been inserted.

But due to the charge redistribution at metal surfaces, it is doubtful whether the *Morse* potential, as well as other phenomenological potentials, can describe surface properties or nanostructures. Here, the *pseudo potential theory* seems to be a much better approach.

2.5.3 *Lennard–Jones Potentials*

The general form of the *Lennard–Jones potential* is [Torrens, 1972]:

$$\upsilon(r) = \frac{\lambda_n}{r^n} - \frac{\lambda_m}{r^m} . \tag{2.22}$$

Originally, it was developed to treat noble gases, but it is often used to describe metals and other forms of solids and liquids. The most commonly form, however, is the so-called Lennard–Jones (6–12) potential, with $n = 12$

Table 2.2 Morse potential parameter set for different metals. The units are 10^{-7} kg, 10^{-21} J, and 10^{-10} m.

Symbol	Mass	ε	ε_c	r_m	α
Al	44.806	43.31	12.881	3.253	1.1646
Ni	97.464	67.37	23.829	2.780	1.4199
Cu	105.52	54.94	18.721	2.866	1.3588
Ag	179.13	53.24	17.034	3.115	1.3690
Pb	344.07	37.62	8.289	3.733	1.1836
Ca	66.553	26.00	15.338	4.569	0.80535
Sr	145.50	24.24	14.372	4.998	0.73776
Cr	86.343	70.72	29.655	2.754	1.5721
Fe	92.735	66.88	24.975	2.845	1.3885
Mo	159.32	128.69	39.802	2.976	1.5079
W	305.28	158.71	50.847	3.032	1.4116
Na	38.177	10.15	13.392	5.336	0.58993
K	64.925	8.690	11.873	6.369	0.49767
Rb	141.93	7.441	4.306	7.207	0.42981
Cs	220.71	7.186	4.165	7.557	0.41569
Ba	228.05	22.69	13.626	5.373	0.65698

and $m = 6$:

$$v(r) = 4\,\varepsilon \left[\left(\frac{\sigma}{r}\right)^{12} - \left(\frac{\sigma}{r}\right)^{6} \right] . \tag{2.23}$$

This potential has its minimum $v_{\min} = -\varepsilon$ at a distance $r = 2^{1/6}\sigma$. Due to its simple form — there are just two parameters — it is often used to describe the cross-interaction of two different materials. Therefore, the interaction potentials of materials a and b are first fitted to *Lennard–Jones* potentials, then the cross-interaction *Lennard–Jones* parameters ε_{ab} and σ_{ab} can be calculated using the *Lorenz–Berthelot* mixing rules

$$\varepsilon_{ab} = \sqrt{\varepsilon_a \varepsilon_b}\,, \quad \sigma_{ab} = \frac{\sigma_a + \sigma_b}{2}\,, \tag{2.24}$$

where ε_a, σ_a and ε_b, σ_b are the *Lennard–Jones* parameters for interactions occurring within materials a and b, respectively [Alvarez *et. al*, 1995; Komvopoulos, Yan, 1997; Dawid, Gburski, 1998]. Cross-interactions can be treated by applying the mixing rules on *Morse* potentials too. Then the mean values for r_m and α have to be used.

Table 2.3 shows a compilation of *Lennard–Jones* parameters for noble gases, copper and silver. While the properties of the noble gases can be

Table 2.3 Lennard–Jones potential parameter set for different materials. The parameters for Argon are from [Verlet, 1967], those for the other noble gases are from [della Valle, Venuti, 1998], and those for metals are from [Halicioglu, Pound, 1975]. The units are 10^{-7} kg, 10^{-21} J, and 10^{-10} m.

Symbol	Mass	ε	σ
Ne	33.51	0.5315	2.786
Ar	66.34	1.6539	3.405
Kr	139.16	2.2075	3.639
Xe	218.02	3.0497	3.962
Cu	105.52	65.626	2.338
Ag	179.13	55.276	2.644

relatively well derived with these potentials (e.g. the accuracy of the melting points varies around 10%) one cannot expect to *Lennard–Jones* potentials describe metallic systems adequately (see notes of the preceding section).

2.5.4 *Barker Potentials for Krypton and Xenon*

Barker [Barker *et. al*, 1974] determined potentials for ground-state krypton–krypton and xenon–xenon interactions, which are as near as possible consistent with a wide range of experimental data including second virial coefficients, gas transport properties, solid-state data, known long-range interactions, spectroscopic information on dimers and measurements of differential scattering cross sections.

While the overlap-dependent many-body interactions have been neglected, the third-order triple dipole three-body interactions have been included. Therefore, the *Barker* potentials must be considered as *effective pair potentials*. Due to the fits to experimental data the analytical form of these potentials is rather complex:

$$\upsilon(r) = \varepsilon[\upsilon_0(r) + \upsilon_1(r) + \upsilon_2(r)] \,, \tag{2.25}$$

where

$$\upsilon_0 = e^{\alpha\left(1-\frac{r}{r_m}\right)} \sum_{i=0}^{5} A_i \left(\frac{r}{r_m} - 1\right)^i - \sum_{i=0}^{2} \frac{C_{2i+6}}{\left(\frac{r}{r_m}\right)^{2i+6} + \delta} \,, \tag{2.26}$$

$$v_1(r) = \begin{cases} e^{\beta\left(1-\frac{r}{r_m}\right)}\left[P\left(\frac{r}{r_m}-1\right)^4 + Q\left(\frac{r}{r_m}-1\right)^5\right] & r \geq r_m \\ 0 & r < r_m \end{cases} , \quad (2.27)$$

$$v_2(r) = \begin{cases} e^{\gamma\left(1-\frac{r}{r_m}\right)^2}\left[R\left(\frac{r}{r_m}-1\right)^2 + S\left(\frac{r}{r_m}-1\right)^3\right] & r \geq r_m \\ 0 & r < r_m \end{cases} . \quad (2.28)$$

Again, ε is the depth of the potential at its minimum, where the value of the inter-nuclear distance is $r = r_m$. Parameters for krypton and xenon are listed in Table 2.4.

In Fig. 2.1, the *Barker* potential for krypton is compared with the corresponding *Lennard–Jones* and *Buckingham* potential. While the *Lennard–Jones* matches nearly perfect those of *Buckingham*, the *Barker* potential shows a deeper minimum. In [Schommers, 1986] the effect of this difference on the structure and surface properties of krypton has been calculated and discussed.

Table 2.4 Barker potential parameter set for krypton and xenon. The units are 10^{-7} kg and 10^{-10} m.

Parameter	Krypton	Xenon
ε	2.787	3.898
r_m	4.0067	4.3623
α	12.5	12.5
A_0	0.23526	0.2402
A_1	−4.78686	−4.8169
A_2	−9.2	−10.9
A_3	−8	−25
A_4	−30	−50.7
A_5	−205.8	−200
C_6	1.0632	1.0544
C_8	0.1701	0.1660
C_{10}	0.0143	0.0323
δ	0.01	0.01
β	12.5	12.5
P	−9	59.3
Q	68.67	71.1
γ	0	−50
R	0	2.08
S	0	−6.24

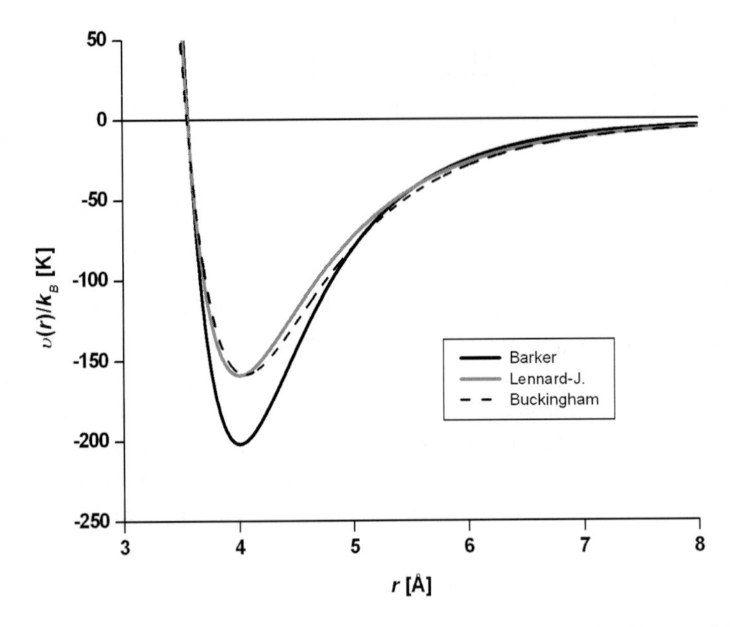

Fig. 2.1 A comparison between the Lennard–Jones, modified Buckingham and Barker potentials for krypton. The parameters are taken from Tables 2.1, 2.3 and 2.4.

In conclusion, of all the mentioned pair potentials in the present chapter the *Barker* potentials are the most accurate ones with the broadest range of validity. Especially the independence of temperature predestines these functions as model potentials within molecular dynamics calculations for nanosystems.

2.6 Pseudo Potentials

In this section, we have a look on the development of the *pseudo potential* approach, which as a result yields two-body interatomic potentials for simple metals. Due to the complexity of the theory it is not possible within the frame of the present monograph, however, to outline all the calculation details, tricks, and fine-tuning features. The following explanations are closely related to those in [Torrens, 1972], which — as could already be seen by the numerous preceding citations — is an extraordinary useful compilation of all topics in the field of pair potentials.

What does the term *pseudo potential* mean? Basically, in the neighborhood of a metal ion an electron experiences two forces:

(1) The strong *Coulomb* attraction of the bare ion that is opposed by the repulsion due to the operation of the *Pauli principle* for the electrons of the closed shells (core electrons).

(2) In the crystal the accumulation of conduction electrons forms a screening charge which in turn balances the ionic charge.

The net effective interaction, experienced by an electron as a result of the cancellation of the two principal contributions, is quite small and is known as *pseudo potential.*

The starting point of the theory is the small core approximation, i.e. the ion cores do not touch or overlap in the solid, and their wavefunctions are not altered by going from the free atom to the metal. This assumption restricts the metals to which the *pseudo potential theory* can be applied. The small core hypothesis leads directly to the exclusion of the conduction electrons from the region of the core, which in turn enables the use of the *orthogonalized plane waves* (OPW) method to describe the conduction electron–ion system. The conduction electrons are assumed to be described by plane waves — orthogonalized to the core functions — as in the free electron theory:

$$OPW_k = |k\rangle - \sum_\alpha |\alpha\rangle \langle\alpha|k\rangle = \left(1 - \hat{P}\right) |k\rangle \ , \tag{2.29}$$

$$|k\rangle = \Omega^{-\frac{1}{2}} \, e^{i\mathbf{kr}} \ , \tag{2.30}$$

$$|\alpha\rangle = \psi_\alpha(\mathbf{r}) \ , \tag{2.31}$$

$$\hat{P} = \sum_\alpha |\alpha\rangle \langle\alpha| \ , \tag{2.32}$$

where $|k\rangle$ represents the free electron wave function and $|\alpha\rangle$ a core state, Ω being the volume of the metal. Introducing the projection operator $\hat{P}$ leads to simpler notations. Now the conduction band state may be expanded in terms of a general linear combination of OPWs based on the reciprocal lattice vectors $\mathbf{q}$:

$$\psi_k = \sum_\mathbf{q} \alpha_\mathbf{q}(\mathbf{k}) \cdot \left(1 - \hat{P}\right) |\mathbf{k} + \mathbf{q}\rangle \ . \tag{2.33}$$

When the OPWs ψ_k are substituted into the *Schrödinger* equation of the system

$$\hat{H}\psi_k = \left[\hat{T} + V(\mathbf{r})\right]\psi_k = E_k\psi_k \tag{2.34}$$

an equivalent equation results with the strong original potential energy term replaced by the much weaker pseudo potential $\hat{W}$:

$$\left[\hat{T} + \hat{W}\right]\phi_k = E_k\phi_k \ , \tag{2.35}$$

$$\hat{W} = V(\mathbf{r}) + \sum_{\alpha}(E_k - E_\alpha)\left|\alpha\right\rangle\left\langle\alpha\right| = V(\mathbf{r}) + \left(E_k - \hat{H}\right)\hat{P} \ . \tag{2.36}$$

The according pseudo wave function ϕ_k is defined by

$$\phi_k = \sum_{\mathbf{q}}\alpha_{\mathbf{q}}(\mathbf{k})\left|\mathbf{k}+\mathbf{q}\right\rangle \ . \tag{2.37}$$

In the exact theory the *pseudo potential* is an energy-dependent operator, but often a more convenient and reasonable approximation is to substitute a simple non-operator term V_p for $(E_k - \hat{H})\hat{P}$. This is known as local pseudo potential:

$$W(\mathbf{r}) = V(\mathbf{r}) + V_p \ . \tag{2.38}$$

The next step is the division of the pseudo potential of the system (Eq. 2.38) into individual ion pseudo potentials w_i and factorization of its matrix elements into a *form factor* $w(\mathbf{q})$ (independent of the ion positions) and a *structure factor* $S(\mathbf{q})$ depending on the ion positions only: if there are N ions at the positions $\mathbf{r}_i$, $i = 1, \ldots, N$, then an electron at $\mathbf{r}$ will have the pseudo potential energy given by

$$W(\mathbf{r}) = \sum_{i=1}^{N} w\left(\left|\mathbf{r} - \mathbf{r}_i\right|\right) \ , \tag{2.39}$$

and the matrix elements (in the local approximation) are

$$\left\langle\mathbf{k}+\mathbf{q}\right|W(\mathbf{r})\left|\mathbf{k}\right\rangle = W(\mathbf{q}) = S(\mathbf{q})w(\mathbf{q}) \ , \tag{2.40}$$

$$S(\mathbf{q}) = \frac{1}{N}\sum_{j=1}^{N} e^{i\mathbf{q}\mathbf{r}_j} \ , \tag{2.41}$$

$$w(\mathbf{q}) = \frac{N}{\Omega} \int w(r) e^{-i\mathbf{q}\mathbf{r}} d\tau \ . \tag{2.42}$$

This form factor — derived from *Fourier transformation* of the individual ion pseudo potential in Eq. 2.39 — is a simplified one based on a *local pseudo potential*. In the non-local operator form it would be energy dependent and it could be calculated from first principles (*perturbation theory*) for matrix elements between states on the *Fermi surface* ($|\mathbf{k}| = |\mathbf{k}+\mathbf{q}| = k_F$). Known as *OPW form factors* several of these, together with the calculation method, are given in [Harrison, 1966]. In addition, there are different semi-local and local approaches where a certain form for the bare ion potential is assumed and then transformed [Heine, Abarenkov, 1964; Animalu, Heine, 1965; Ashcroft, 1966; Shaw, Harrison, 1967; Shaw, 1968]. Such potentials are known as *model potentials* and are usually of simple form. The final potential, however, is sensitive to the exact form of the core potential only when the separation between core and conduction electron states is small.

The next step after determining the bare ion form factor, which theoretically contains all the information necessary to describe the ion, is the consideration of the screening by the conduction electrons. This is achieved in the linear approximation where the bare ion and screened ion form factors (w_b and w_s) are related by a dielectric function $\varepsilon(q)$:

$$\langle \mathbf{k}+\mathbf{q}|w_s(\mathbf{r})|\mathbf{k}\rangle = \frac{\langle \mathbf{k}+\mathbf{q}|w_b(\mathbf{r})|\mathbf{k}\rangle}{\varepsilon(q)} \tag{2.43}$$

or in the simpler case of the local pseudo potential

$$w_s(q) = \frac{w_b(q)}{\varepsilon(q)} \ . \tag{2.44}$$

For metals — compared to an electron gas — the dielectric formulation is an approximation. Depending on how or whether exchange and correlation between the conduction electrons are included, the dielectric function may take several forms [Heine, Abarenkov, 1964; Hubbard, 1958; Geldart, Vosko, 1966; Sham, 1965; Shaw, Pynn, 1969; Singwi *et. al*, 1970]. An example for a simple but frequently used expression of $\varepsilon(q)$ is the *Hartree dielectric function* for the electron gas without any interactions between the electrons:

$$\varepsilon(q) = 1 + \frac{2k_F m\, e^2}{\pi \hbar^2 q^2} \left(1 + \frac{4k_F^2 - q^2}{4k_F\, q} \ln \left| \frac{2k_F + q}{2k_F - q} \right| \right) \ . \tag{2.45}$$

Here it is important to note that most forms of the dielectric function contain a logarithmic singularity which gives rise to *oscillations* in the real-space potential after the *Fourier transformation*. In addition, the dielectric function may contain one or more parameters adjustable to experiment.

The total energy of the ion–electron system may now be found to the second order using perturbation theory. The energy of the eigenstate of wave number k is given by the equation

$$E_k = \frac{\hbar k^2}{2m} + \langle \mathbf{k}|W_s(\mathbf{r})|\mathbf{k}\rangle + \frac{\hbar^2}{2m}\sum_{q\neq 0} \frac{\langle \mathbf{k}+\mathbf{q}|W_s(\mathbf{r})|\mathbf{k}\rangle \, \langle \mathbf{k}|W_b(\mathbf{r})|\mathbf{k}+\mathbf{q}\rangle}{k^2 - |\mathbf{k}+\mathbf{q}|^2} \, . \tag{2.46}$$

For further considerations, only the third term is relevant since the first and second term do not influence the ion–electron–ion interaction under constant volume conditions (the first term is just the kinetic energy of the electrons if there were no ions present, while the second concerns the ions but assumes that the electrons are completely free). Now the electron energy can be derived by integrating the third term of Eq. 2.46 up to the *Fermi wave number* k_F and by factorizing into structure and form factors. This is known as the *band structure energy* per ion:

$$E_{bs} = \sum_{q\neq 0} S^*(q) S(q) F(q) \, , \tag{2.47}$$

where $F(q)$ is the so-called *energy-wave number characteristic*

$$F(q) = \frac{\Omega}{4\pi^3 N} \frac{\hbar^2}{2m} \int_{|k|<k_F} \frac{\langle \mathbf{k}|W_b(\mathbf{r})|\mathbf{k}+\mathbf{q}\rangle \, \langle \mathbf{k}+\mathbf{q}|W_s(\mathbf{r})|\mathbf{k}\rangle}{k^2 - |\mathbf{k}+\mathbf{q}|^2} \, d^3k \tag{2.48}$$

with

$$k_F = \left(\frac{3\pi^2 N\, Z}{\Omega}\right)^{\frac{1}{3}} \, . \tag{2.49}$$

Finally, the interaction potential between two ions of the ion–electron system may now be divided into the direct *Coulomb* interaction, unaffected by the intervening electrons and most often approximated by point charges with the effective valency Z^*, and the indirect ion–electron–ion interaction:

$$v(r) = v_d(r) + v_{\text{ind}}(r) \, , \tag{2.50}$$

$$v_d(r) = \frac{Z^{*2} e^2}{r} \, . \tag{2.51}$$

The indirect interaction can be obtained directly from the band structure energy. Using the structure factor notation of Eq. 2.41 within Eq. 2.47 leads to

$$E_{bs} = \frac{1}{N^2} \sum_{q \neq 0} \sum_{\mathbf{r}} F(q) e^{-i\mathbf{qr}} = \frac{1}{N^2} \sum_{q \neq 0} F(q) + \frac{1}{N} \sum_{r \neq 0} v_{\text{ind}}(r) \,, \qquad (2.52)$$

where

$$v_{\text{ind}}(r) = \frac{1}{N} \sum_{\mathbf{q},\mathbf{r} \neq 0} F(q) e^{-i\mathbf{qr}} = \frac{\Omega}{\pi^2} \int_0^\infty F(q) \frac{\sin qr}{qr} q^2 dq \,. \qquad (2.53)$$

In conclusion, the validity of this *pseudo potential* is restricted by reason of the basic assumptions to non-overlapping ion cores and constant volume in the system. Further, evidently this potential is useful only in describing the interactions near the equilibrium separation in the crystal as well as in situations where the distribution of conduction electrons in the region of an ion resembles that in the case of a perfect crystal.

2.6.1 *Schommers Potential for Aluminium*

With respect to a higher degree of accuracy and a broader range of validity (surface studies and nanostructure applications) *Schommers* developed a pair potential for aluminium on the basis of *pseudo potential theory* combined with phenomenological approaches as well as *effective* pair potential considerations.

The latter has been introduced into the description of the direct ion interaction (Eq. 2.51) by adding an additional *van-der Waals* type interaction $f(r)$ [Schommers, 1976] to the *Coulomb* term

$$f(r) = -\frac{\alpha_1}{r^6} - 2Z^* \frac{\alpha_2}{r^4} \,. \qquad (2.54)$$

This considers dipole–dipole and monopole–dipole interactions that result from the finite extension of the ion cores.

Within the derivation of the pseudo potential a similar expression $h(r)$ has been added to the bare ion potential $w(r)$ to consider monopole–multipole interactions between the conduction electron and the ion core [Schommers, 1976]:

$$h(r) = -\frac{\alpha_2}{r^4} + \cdots \,, \qquad (2.55)$$

where the first term describes the monopole–dipole interaction.

Further, for the derivation of the screened ion form factor the static *Hartree dielectric function* (Eq. 2.45) has been replaced by a more detailed expression.

While it is a very good approximation to treat the core contributions independent of the temperature, the long-range part of pseudo potentials depend critically on the electronic arrangement which is sensitive to temperature variations, i.e. an interaction potential for metals generally has to be temperature dependent. Therefore, the description of the *Schommers* potential for aluminium has been performed in dependency of the (temperature dependent) material density or lattice constant, respectively [Schommers *et. al*, 1995].

The *Schommers* pair potential for aluminium is shown in Fig. 2.2 for two different temperatures in comparison with the according *Morse* potential from Table 2.2 (with the corrected value ε_c) and an additional *Morse* potential, that is fitted to the *Schommers* potential.

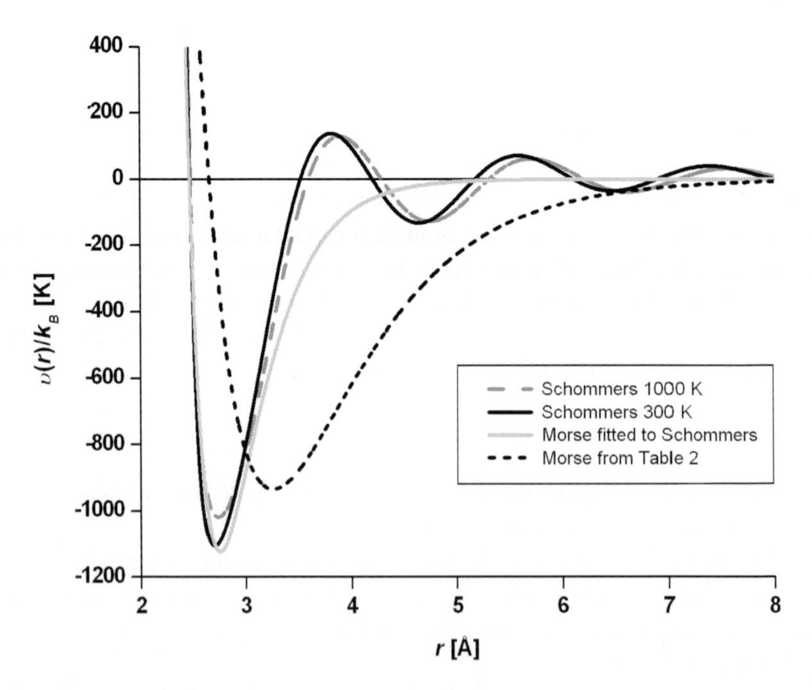

Fig. 2.2 A comparison of the Schommers and Morse pair potential for aluminium. The Schommers potential is plotted for 300 K and 1000 K. The parameters of the Morse potential fitted to the Schommers potential are $\varepsilon = 14.5$, $r_m = 2.75$, $\alpha = 2.5$ in the same units as given in Table 2.2. The melting temperature of aluminium is 933 K.

The differences can be seen quite clearly: beside the deviations in depth and location of the minimum the *Morse* potential completely lacks the long-range oscillations due to the ion–electron–ion interactions in the metallic crystal. This is the reason why the use of phenomenological potentials, such as those from *Morse*, *Lennard–Jones* or *Buckingham*, for the description of metallic interactions is rather questionable.

Due to the additional fitting of free parameters of the *Schommers* potential to numerous experimental data it describes a lot of basic aluminium properties accurately within a wide temperature range [Schommers *et. al*, 1995; Rieth *et. al*, 1999], e.g. melting point, diffusion constant of the liquid phase, structure of solid and liquid bulk, phonon density of state, mean square displacements at the surface, onset of pre-melting, etc.

Since all of these are critical properties with respect to the behavior of nanosystems and since the *Schommers* potential for aluminium is able to describe them with sufficient accuracy and dependent on temperature, this potential has been chosen for most of the molecular dynamic studies presented within the present monograph.

2.7 Many-Body Potentials

Though many-body potentials are of minor relevance for the present work — since they have to be used mainly for the description of interactions within covalent bonded materials or for phenomena at higher energy levels — we would like to outline at least some common approaches.

Allinger developed widely used models (MM2, MM3) for a broad range of organic structures [Burkert, Allinger, 1982; Allinger *et. al*, 1989] which are based on the separation of the total interaction between particles, that are significant for a bond, into a sum of single potentials with respect to bond stretching, bond angle bending, bond torsion and non-bonded interaction. The latter is described in terms of *modified Buckingham potentials* with parameters listed in Table 2.1. Bond stretching is treated with cubic pair potentials, whereas bond angle bending is modeled in form of sextic three-body potentials dependent on the angle between two bonds to a shared atom. Finally, the description of bond torsion follows an expression depending on the torsion angle between two bonds within the same plane and a third one, thus acting as a four-body potential.

For the development of such models a lot of *a priori* knowledge is necessary to classify all the different bond types and to treat the special cases.

The MM2 and MM3 (*molecular mechanics*) models are used to characterize the minimum energy configurations of structures, their stiffness, bearing properties and the like [Drexler, 1992], and they have become standards in the chemical literature.

Due to their complexity and difficult derivation, there are basically just a few different three-body potentials available which are well suited for molecular dynamics calculations. A comparative study and discussion of some of these empirical potentials, which have been applied mainly to carbon, silicon and other semiconductor materials, is given in [Balamane, 1992].

Another special type of potential that describes metallic interactions is derived from the *Embedded Atom Method* (*EAM*) [Daw, Baskes 1983]. It is a generalization of the *quasi-atom theory* [Stott, Zaremba, 1980] that treats all atoms in a unified way. Its name comes from the fact, that it views each atom as embedded in a host lattice consisting of all other atoms. Here, the total energy of a system with N atoms is given by

$$E_{\text{tot}} = \sum_{i=1}^{N} \left[F_i(\bar{\varrho}_i) + \frac{1}{2} \sum_{\substack{j=1 \\ j \neq i}}^{N} u_{ij}(r_{ij}) \right] , \tag{2.56}$$

where $F_i(\bar{\varrho}_i)$ is the energy required to embed atom i into the background electron density $\bar{\varrho}_i$ at site i, and $u_{ij}(r_{ij})$ is the core–core pair interaction potential between atoms i and j separated by the distance r_{ij}.

The host electron density $\bar{\varrho}_i$ is approximated by a linear superposition of the spherically averaged electron densities of the n_i atoms neighboring atom i:

$$\bar{\varrho}_i = \sum_{\substack{j=1 \\ j \neq i}}^{n_i} \varrho(r_{ij}) , \tag{2.57}$$

where $\varrho(r_{ij})$ is the electron density of atom j at a distance r_{ij} from the nucleus of atom i. If the atomic densities $\varrho(r)$ and the pair interaction $u(r)$ are both known, the embedding energy F can be uniquely defined by matching a certain equation for the cohesive energy of the metal as a function of the lattice constant. This can be used to fit $F(\bar{\varrho})$ to experimental results. Different approaches and fits to the fcc metals — Ni, Pd, Cu, Ag, and Au can be found, for example in [Foiles *et. al*, 1986; Voter, Chen, 1987; Rey *et. al*, 1993].

In contrast to pair potentials which depend on the interatomic distance only, the according EAM potentials have an additional term that depends on the background electron density. This in turn is defined by the positions of the n_i next neighbor atoms (Eq. 2.57) which act indirectly as n-body contributions.

Chapter 3

Molecular Dynamics

As has already been shown, the problem of *ab initio* methods consists in the derivation of the potential energy surface (Eq. 2.12) and finding its local minima dependent on the $3N$ co-ordinates of the N nuclei. Using the pair potential approximation or even many-body potentials with the *molecular mechanics* method significantly reduces the effort for obtaining the potential energy surface, while the problem of detecting the saddle points still remains the same. Though there are a lot of different strategies available, there are no (or at least very few) systematic algorithms that localize the energetically best fitting configurations within an acceptable time.

However, while *molecular mechanics* as well as *ab initio* methods are very popular and successful in the description of more or less complex systems (molecules, clusters, etc.) — most often known from organic chemistry — the treatment of the many-particle problem is handled, nevertheless, in a static way, i.e. the results are only valid for temperatures equal to zero. Even by including temperature dependent potentials *molecular mechanics* calculations can never deliver dynamic values. But especially with respect to nanostructures consisting of non-bonded materials or metals, such dynamic effects are of considerable interest.

Since the *harmonic approximation* for crystalline solids — known from *solid state physics* — is not well suited for an adequate description of surface phenomena at higher temperatures either [Schommers, 1986], the only method at disposal is the classical solution of *Hamilton's equations of motion* for N particles

$$\dot{\mathbf{q}}_i = \frac{\partial H}{\partial \mathbf{p}_i} \ , \ \ \dot{\mathbf{p}}_i = -\frac{\partial H}{\partial \mathbf{q}_i} \ , \ \ i = 1, \cdots, N \ , \tag{3.1}$$

where $\mathbf{q}_i$ and $\mathbf{p}_i$ are a set of generalized co ordinates and momenta.

There are two different ways for the calculation of the according forces: one can choose either the potential energy surface determined by quantum mechanics, or one uses pair (or many-body) potentials.

The first method — known as *quantum molecular dynamics* [Seifert, 1998] — is involved with the same problems already discussed in connection with the *ab initio* methods, i.e. it is only applicable with the restriction to a very small number of atoms, whereas the limitation of the latter — the "classical" *molecular dynamics* (MD) — is related to the validity of the pair potential approximation.

In the frame of this monograph, we consider nanostructures consisting of either krypton or aluminium atoms, which interact through effective pair potentials without any further outer forces. In this case the *Hamilton* function in *Cartesian* co-ordinates $r = [x, y, z]$ reads

$$H(\mathbf{p}_i, \mathbf{r}_i) = \sum_{i=1}^{N} \frac{\mathbf{p}_i^2}{2m_i} + U(\mathbf{r}_1, \cdots, \mathbf{r}_N) \tag{3.2}$$

from which we obtain *Newton's equations of motion*

$$\dot{\mathbf{r}}_i = \mathbf{v}_i \ , \tag{3.3}$$

$$\dot{\mathbf{v}}_i = \ddot{\mathbf{r}}_i = \frac{\mathbf{F}_i}{m_i} \ , \tag{3.4}$$

where $\mathbf{F}_i = [F_x, F_y, F_z]_i$ is the force acting on the ith atom of mass m_i and $\mathbf{v}_i = [v_x, v_y, v_z]_i$ is its velocity according to $\mathbf{v}_i = \mathbf{p}_i/m_i$. The forces are derived from the potential U as follows

$$\mathbf{F}_i = -\frac{\partial}{\partial \mathbf{r}_i} U(\mathbf{r}_1, \cdots, \mathbf{r}_N) \ , \tag{3.5}$$

and since U corresponds to the pair potential v according to Eq. 2.15 (to be more specific, for krypton we use the *Barker* potential with Eqs. 2.25–2.28 and for aluminium the *Schommers* potential as plotted in Fig. 2.2), we can express the magnitude of the interacting force f_{ij}

$$f_{ij} = |\mathbf{F}_{ij}| = -\left. \frac{\partial}{\partial r} v(r) \right|_{r=r_{ij}} \ , \tag{3.6}$$

with r_{ij} being the distance between two atoms i and j

$$r_{ij} = |\mathbf{r}_i - \mathbf{r}_j| = \sqrt{(x_i - x_j)^2 + (y_i - y_j)^2 + (z_i - z_j)^2} \ . \tag{3.7}$$

With this the total force for particle i can be easily obtained by the expression

$$\mathbf{F}_i = \sum_{\substack{j=1 \\ j \neq i}}^{N} \frac{\mathbf{r}_i - \mathbf{r}_j}{r_{ij}} \frac{\partial}{\partial r} v(r) \bigg|_{r=r_{ij}} \tag{3.8}$$

$$= \sum_{\substack{j=1 \\ j \neq i}}^{N} \frac{(\mathbf{r}_j - \mathbf{r}_i) f_{ij}}{\sqrt{(x_i - x_j)^2 + (y_i - y_j)^2 + (z_i - z_j)^2}} \,.$$

After setting the initial values — positions and velocities — for all atoms, with help of Eq. 3.8, first, the forces are calculated, then the set of first-order differential equations (Eqs. 3.3 and 3.4) is solved numerically for a certain time step with the new positions and velocities — which are saved — as a result. Then the new forces are obtained to solve the equations of motion for the next time step, and so on.

Basically, *molecular dynamics* is as simple as that. But, going into detail, the implementation of a state-of-the-art MD software includes a lot of further considerations which are connected to the field of *numerics* as well as *computer science*. Due to the large number of different applications related to nanostructured systems, an universal MD software is not available and, therefore, such topics as model formation, integration and differentiation algorithms, effective force interaction computation and graphical presentation have to be discussed in order to assemble an optimal program package for each specific case.

3.1 Models for Molecular Dynamics Calculations

If the assumption of two-body interactions leads to a reasonable description of specific system properties, MD calculations can be considered as perfect computer experiments, since the total information of the system (position and velocity of all particles) is available for each time step. From this, further characteristics can be derived without any additional parameters.

While real experiments are most often complex, time consuming and therefore very expensive, it is reasonable to simulate such experimental studies. Moreover, it is rather tempting to verify ideas or to perform hypothetical experiments that in reality still cannot be carried out, but might possibly gain some relevance in the near future.

Depending on the system, there has to be built more or less complicated models as an input to MD calculations. In the case of (metallic) nanosystems, it is easy to imagine many different scenarios as, for example, interactions of nanostructures with surfaces (liquid or solid) or nano-machines, both in combination with specific environments depending on temperature, pressure, mechanical load, vacuum or gas atmosphere, to name just a few.

3.1.1 *Initial Values*

The first step of setting up a model is to define the initial position and velocity for each atom. In the case of liquids or gases, the distribution of the particles can be chosen randomly with the appropriate density. Sometimes this may lead to substantial overlaps. In these cases, a minimum value for the interatomic distance has to be set and verified.

In the case of crystals, the initial positions are given by the perfect lattice structure according to the system under investigation. Here, aluminium as well as krypton show a face-centered cubic (fcc) lattice (the lattice constant L is 4.03 Å for Al at 50 K and 5.73 Å for Kr at 70 K).

It is relatively easy to accomplish clusters, structured in the shape of a cuboid, either by assembling unit cells in rows and columns or by piling up one layer after the other. When surfaces are of interest the co-ordinates of the unit cell have to be transformed into the according orientation. The geometry for the construction of fcc crystals with (001), (011) and (111) surfaces are given in Figs. 3.1–3.3.

The initial velocities for the particles of resting objects have to meet the conservation law of momentum, i.e. the velocities have to be chosen so that there is no overall momentum

$$\mathbf{P} = \sum_{i=1}^{N} m_i \mathbf{v}_i = \mathbf{0} \,. \tag{3.9}$$

In thermal equilibrium, the velocities would be distributed according to the *Maxwell* distribution.

Since most programming languages do only support *uniformly* distributed random numbers but no *Maxwell* distributions, it is easier, faster and, however, sufficient to start the MD calculations with a simple initial velocity distribution and to continue until equilibrium is established due to the particle collisions.

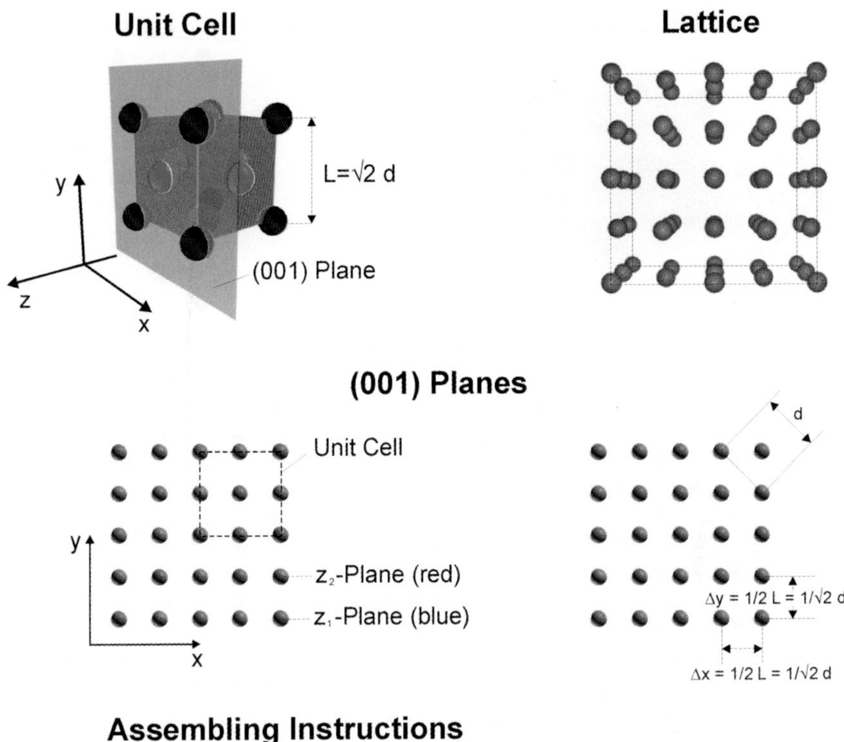

Fig. 3.1 Geometry of (001) planes for face-centered cubic crystals. Cuboids with (001) surfaces can be assembled layer by layer.

A simple possibility is to choose the same magnitude v_m of the velocities for all particles according to

$$|\mathbf{v}_i| = v_m = \sqrt{\frac{3k_BT}{m_i}} \tag{3.10}$$

where m_i is the mass of particle i, k_B is the *Boltzmann* constant and T is the kinetic temperature.

After the determination of the magnitudes of the particle velocities, one still needs the *uniformly* distributed velocity directions, which can be produced by a number of methods. The simplest method for generating

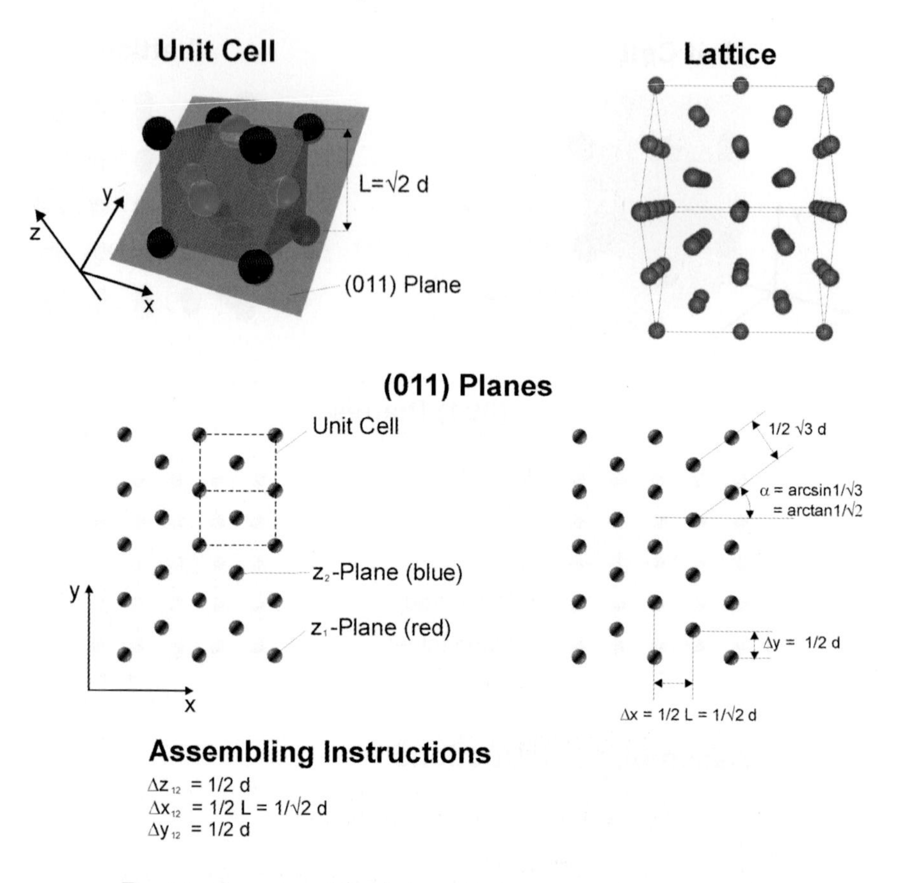

Fig. 3.2 Geometry of (011) planes for face-centered cubic crystals.

a random (unit) vector $\mathbf{e}_u$ on the surface of a sphere is the acceptance–rejection technique of [von Neumann, 1951] with

$$\left.\begin{aligned} \mathbf{e}_u &= \tfrac{1}{n}\left[1-2n_1, 1-2n_2, 1-2n_3\right] \\ n &= \sqrt{n_1^2+n_2^2+n_3^2}\,, \quad n<1 \end{aligned}\right\} \tag{3.11}$$

or an improved version [Allen, Tildesley, 1990; Marsaglia, 1972] with

$$\left.\begin{aligned} \mathbf{e}_u &= \left[2(1-2n_1)\sqrt{1-n^2}, 2(1-2n_2)\sqrt{1-n^2}, 1-2n^2\right] \\ n &= \sqrt{n_1^2+n_2^2}\,, \quad n<1 \end{aligned}\right\} \tag{3.12}$$

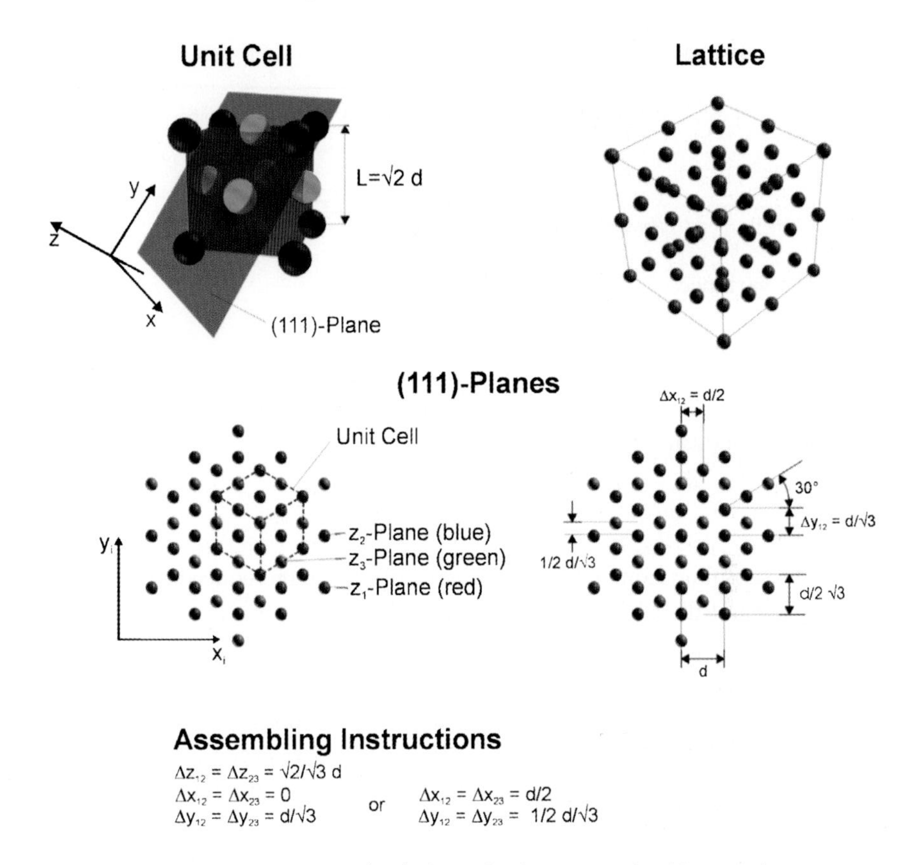

Fig. 3.3 Geometry of (111) planes for face-centered cubic crystals.

where n_1, n_2, n_3 are *uniformly* distributed values in the range $[0, 1]$ which can be easily obtained from any common random number generator. If n is smaller than one, the unit vector is accepted, otherwise the procedure is repeated.

Now, the easiest way to ensure the conservation law (Eq. 3.9) is to generate a velocity vector $\mathbf{v}_i = |\mathbf{v}_i|\mathbf{e}_u$ as outlined above and set the initial velocity vector and its opposite $(-\mathbf{v}_i)$, respectively, for a pair of particles. If the model contains an odd number of particles the initial velocity of the remaining particle is simply set to zero.

Another possibility is to choose the components v_x, v_y, v_z of the velocity vectors $\mathbf{v}_i = [v_x, v_y, v_z]_i$ randomly from the *Gaussian* parts $\rho_{x,y,z}$ of

Maxwell's distribution ρ_M

$$\rho_M(\mathbf{v}) = \rho_x(v_x)\rho_y(v_y)\rho_z(v_z) \,, \tag{3.13}$$

where

$$\rho_{x,y,z}(v_{x,y,z}) = \frac{1}{\sqrt{2\pi\sigma^2}} e^{-\frac{(v_{x,y,z}-v_{cm})^2}{2\sigma^2}} \tag{3.14}$$

with the variance

$$\sigma^2 = \frac{k_B T}{m_i} \tag{3.15}$$

and the mean velocity component (in one dimension) v_{cm} according to the kinetic temperature T

$$v_{cm} = \sqrt{\frac{k_B T}{m_i}} \,. \tag{3.16}$$

Now, sampling from a *Gaussian* distribution can be reduced to sampling from a *normal* distribution (zero mean and unit variance):

$$v_{x,y,z} = v_{cm} + \sigma x_n \,, \tag{3.17}$$

where $v_{x,y,z}$ are the desired vector components according to the *Gaussian* distributions $\rho_{x,y,z}$ (Eq. 3.14) and x_n is a *normal* distributed value, which in turn can be obtained from a common (*uniform*) random number generator either by [Box, Muller, 1958]

$$x_n = \sqrt{-2\ln n_1}\cos 2\pi n_2 \,, \tag{3.18}$$

or

$$x_n = \sqrt{-2\ln n_2}\sin 2\pi n_1 \,, \tag{3.19}$$

or approximately by [Allen, Tildesley, 1990]

$$x_n = \sum_{k=1}^{12} n_k - 6 \,, \tag{3.20}$$

where $n_1 \cdots n_{12}$ are *uniformly* distributed values in the range $[0,1]$.

In conclusion, derived from random numbers, the initial velocities could be set approximately according to the thermal equilibrium (*Maxwell distribution*), whereas for a large number of particles the overall momentum (Eq. 3.9) should nearly be zero. For smaller particle numbers, however, it is better to cancel the momenta pairwise, as mentioned above.

3.1.2 *Isothermal Equilibration*

Independent of the methods chosen to set the initial values the system is not exactly in equilibrium, therefore, the model has to be equilibrated simply by performing MD calculations for a sufficient number of time steps. Usually the equilibrium is established very quickly, i.e. after several hundred calculation steps at most. But how can this be verified?

For this an useful function is given in [Schommers, 1986]. With help of the velocity vectors $\mathbf{v}_i$ obtained from the MD calculation for the N particles

$$\alpha(t) = \frac{\left\langle \mathbf{v}^2(t)\mathbf{v}^2(t)\right\rangle}{\left\langle \mathbf{v}^2(t)\right\rangle \left\langle \mathbf{v}^2(t)\right\rangle} = \frac{\frac{1}{N}\sum_{i=1}^{N}\left[\mathbf{v}_i^2(t)\right]^2}{\left[\frac{1}{N}\sum_{i=1}^{N}\mathbf{v}_i^2(t)\right]^2} \tag{3.21}$$

gives a time-dependent measure for the degree of distribution. If the magnitudes of all particle velocities are the same, $\alpha(t) = 1$, for *Maxwell* distributed velocities — and therefore in thermal equilibrium — $\alpha(t)$ takes the value of $5/3$. As can be seen from Fig. 3.4, the equilibration process takes a relatively short time and after that $\alpha(t)$ fluctuates around the equilibrium

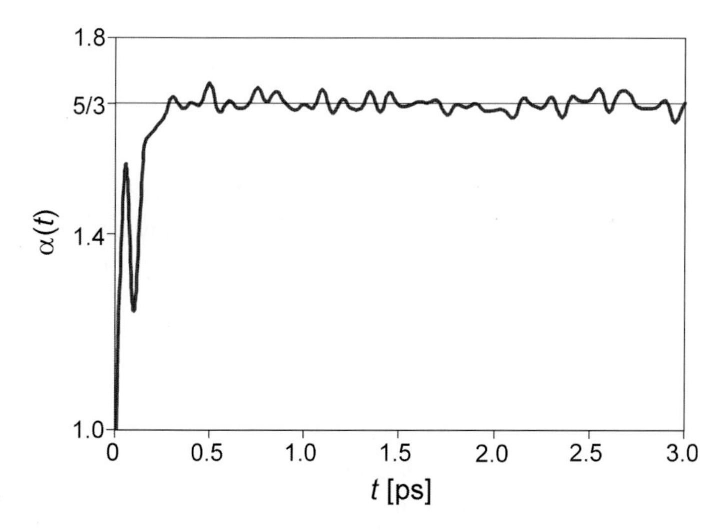

Fig. 3.4 The velocity distribution of a cubic nano-cluster consisting of 4000 aluminium atoms characterized by means of the function $\alpha(t)$. The positions have been set according to the perfect crystal lattice with (001) surfaces and the magnitudes of the initial velocities are the same for all atoms, therefore $\alpha(t)$ starts with the value 1. After 0.3 ps (300 time steps) the model is in thermal equilibrium (Maxwell distribution of the velocities) and due to the finite number of atoms $\alpha(t)$ fluctuates around the according value of $5/3$.

value of 5/3, whereas this fluctuations are due to the finite number of particles.

The temperature of the model expressed by the mean-square velocities of the N particles is given by

$$T(t) = \frac{1}{N} \frac{m_i}{3k_B} \sum_{i=1}^{N} \mathbf{v}_i^2(t) \tag{3.22}$$

and plotted as a function of time shows a similar behavior as $\alpha(t)$, i.e. the system temperature starts with a value specified by the initial velocities (Eqs. 3.10 or 3.15 and 3.16) and frequently changes until equilibrium is established.

But most often the equilibrium temperature is an important system parameter, according to which the model has to be designed, that is, one wants to prescribe a certain value for the equilibrium temperature in advance. Though a crude but sufficient way, this can be done by re-scaling the velocities according to

$$\mathbf{v}_i(t) := \sqrt{\frac{T_d}{T}}\ \mathbf{v}_i(t)\ , \quad i = 1, \cdots, N\ , \tag{3.23}$$

where T_d is the desired model temperature, and T and $\mathbf{v}_i$ are the current temperature and velocity vectors. During such an isothermal equilibration phase this method needs to be applied after about each tenth time step only.

The re-scaling of velocities is not only useful for the equilibration of a MD model, but can be applied during MD calculations too, where a variation of T_d with time leads to artificial tempering or cooling processes. In this way, it is possible to determine stable configurations of clusters as an alternative to the *molecular mechanics* methods [Kirkpatrick *et. al*, 1983].

It should be noted that there are more elegant methods to keep the temperature constant (e.g. [Andersen, 1980; Andersen *et. al*, 1984]). Most of them use an additional velocity-dependent term in the equations of motion to prescribe the system temperature. This is called *isothermal* or *constant NVT molecular dynamics* (or the use of *canonical ensembles* in terms of *statistical mechanics*). But for the equilibration phase the simple re-scaling of the velocities according to Eq. 3.23 is sufficient.

3.1.3 *Boundaries*

When performing MD calculations with free nano-clusters of moderate temperature there is no need for a limitation of the simulation space. But as soon as liquid or gaseous fractions are involved the whole system expands into all directions with increasing time. Therefore, sometimes it may be appropriate to limit the simulation area by introducing spatial boundaries.

The simplest case of a boundary is a reflecting box. Whenever a particle reaches the surface of this box during the MD calculations its perpendicular velocity component is reverted. By applying such a *boundary condition* the energy is conserved and the volume as well as the particle number is kept constant (isolated system, *micro-canonical ensemble*).

Another application, for instance, could be to make the reflection dependent on the particle momentum, i.e. if the perpendicular velocity exceeds a certain value it can leave the box and is removed from the MD calculation, or even more elegantly, the box is replaced by a spherical potential barrier (in addition to the interatomic interactions) which a particle can overcome only with sufficient kinetic energy. Such a model could be applied, for example, on the simulation of membranes, where the boundary potential acts as the so-called *chemical potential* (*grand canonical ensemble*).

Since boundary conditions define the thermodynamic environment, they are a substantial part of MD models. Here, the application of *periodic boundary conditions* are one of the most important techniques. The idea arose from the question of how to simulate, for instance, the crystal or liquid bulk with just a few hundred or thousand particles, while in nature such structures are built of several 10^{23} atoms at least.

The trick is again to restrict the simulation space to a cubical box, that the particles cannot leave. Then — instead of total reflection — a particle that exits the box from one face re-enters the box from its opposite face without changing the velocity vector as is illustrated in Fig. 3.5.

Such a treatment of the particle positions ensures the constancy of energy, particle number and volume. But the computation of the particle interactions has to be expanded according to these boundary conditions.

Therefore, the boundary box including all particles (simulation space) has to be surrounded by 26 virtual copies (see Fig. 3.6). Then the interatomic forces are computed not only for the atoms within the simulation space, but by including all the particles of its virtual images. These images can be easily created by adding or subtracting the according box length from the co-ordinates of the particle position vectors.

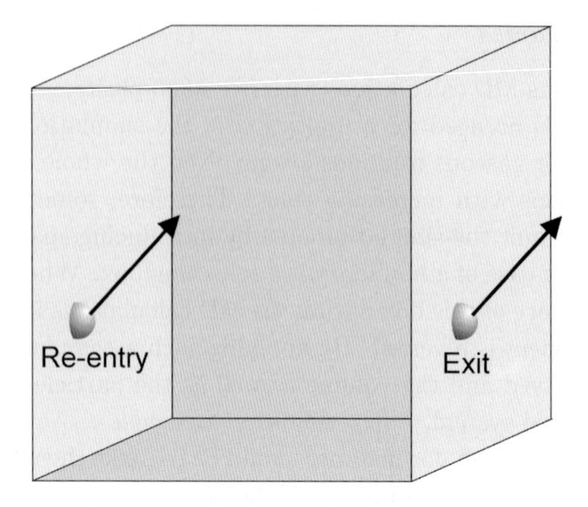

Fig. 3.5 Periodic boundary conditions: When a particle is going to leave the simulation box its position is set immediately to the opposite face and the velocity vector remains unchanged.

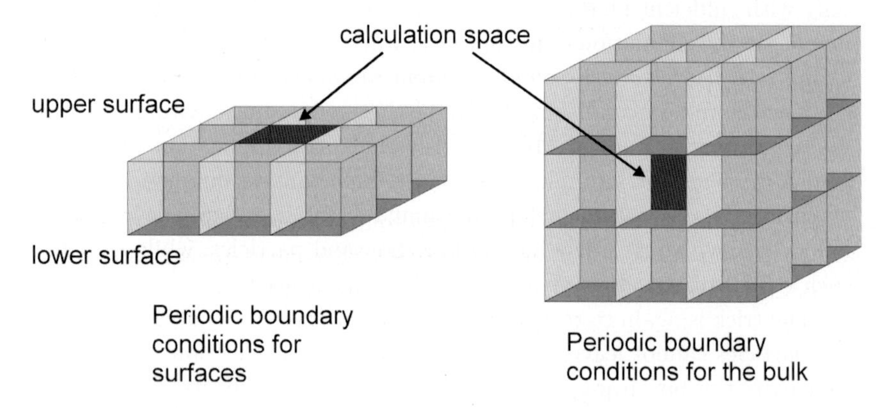

Fig. 3.6 Periodic boundary conditions: In order to simulate an infinite system the computation of particle interactions is extended to virtual copies of the calculation space. For bulk models it needs 26 of these images to cover all directions. To model free surfaces the calculation box is surrounded by 8 of its images.

Due to its periodic repetitions into all directions such a model simulates an infinite extended system (bulk) on the basis of a finite number of particles.

Another important MD model is that for free surfaces. This can be derived from the model for the bulk simply by applying the periodic boundary conditions in two dimensions only (Fig. 3.6).

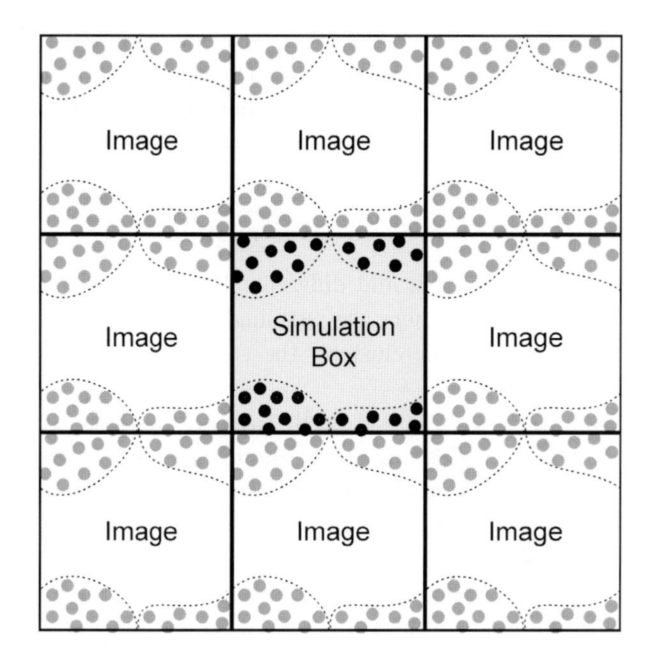

Fig. 3.7 A two-dimensional illustration of the effect of periodic boundary conditions. The positions of the particles of the simulation box (black circles) are mirrored into all directions to compute the interactions. Therefore, the resulting structures (in this case) are rows of clusters (enclosed in dotted lines).

The application of periodic boundary conditions (the effect is illustrated in Fig. 3.7) implies a further restriction with respect to the interaction potential. If the range of the potential was longer than the box length, a particle would interact with itself. Since such nonsense must be avoided, it is necessary to limit the range of the potentials.

This is only one of several reasons to introduce the so-called *cut-off radius* r_c which is the maximum interaction distance. The easiest method is literally to cut the potential function off, i.e. for distances greater than r_c the interaction is set to zero.

A more elegant approach is the use of a cut-off function that pushes the values of the potential or force function smoothly to zero, like for example

$$f_c(r) = \begin{cases} 1 & \text{for} \quad r \leq r_{sc} \\ \frac{1}{2}\left(1 + \cos\frac{r - r_{cs}}{r_c - r_{cs}}\pi\right) & \text{for} \quad r_{cs} < r < r_c \\ 0 & \text{for} \quad r \geq r_c \end{cases}, \tag{3.24}$$

where r_{cs} denotes the onset of the fitting zone.

Further, in the case of crystals one has to pay attention to the initial positions of the atoms. The positions have to be set according to the perfect lattice, but with respect to a continuous repetition at the boundaries (an example that illustrates the continuation effect is given in Fig. 3.7).

3.1.4 *Nano-Design and Nano-Construction*

So far the general procedures and different elements of MD models have been discussed. As long as there are just single and simple shaped objects of interest, like cubical nano-clusters, the set-up of an appropriate MD model is relatively easy.

Since the initial positions are given by the perfect lattice structure such objects can be built by using any programming language to implement the according algorithms. Depending on the variety of materials, characterized by the lattice constant and structure (fcc, bcc, hexagonal, etc.), and depending on the possible orientations ((001), (011), (111) and so on) the software development is more or less extensive. Further, for the generation of surface and bulk models the additional treatment of periodic boundary conditions is necessary.

But obviously, such simple objects are only of minor interest within the field of *nano-engineering*, where either structures with complex shapes or interactions (cluster–cluster or surface-cluster) under various conditions play an important role. Especially for studies of *nano-machines* (but for others too) the MD models have to contain moving parts in addition to objects of miscellaneous materials, and therefore, the software should be able to combine single structures, each with different initial values.

Quite clearly, the development of a computer program that can generate MD models for all the above mentioned topics is rather extensive. While usually there is no way to avoid programming at least for simple structures (cuboids, spheres, surfaces), for the set-up of the initial values according to rotational or transitional movements and for the construction (positioning, centering, adjusting) of individual objects, the *nano-designing* part may be accomplished by a common ploy: the "abuse" of commercial software.

The idea arose from the fact that for metallic nanostructures — unlike covalent bonded materials — there are basically no restrictions concerning their initial shapes, i.e. in this case a MD model can be built first by defining the outer shape and then by filling it up with atoms according to the crystal lattice. Since the design of three-dimensional shapes is a very common topic within *mechanical engineering* — the so-called *computer aided design* (*CAD*) — one can use a suitable CAD software with a programming interface to perform such a task.

For the present work the choice fell on the commercial software package *Genius Desktop* [Autodesk, 1998] (in combination with *Mechanical Desktop* [Autodesk, 1998]) that includes a LISP derivative [Autodesk, 1998] as programmable interface. At first, two additional system functions have been developed, one to compute the outer limits of a 3d–CAD object, the other to determine whether a given spatial point lays within a selected object or not. Then, with the help of these functions, the implementation of various filling, scaling and storage algorithms for different lattice structures has been performed. An example for such a *nano-design* is illustrated in Fig. 3.8.

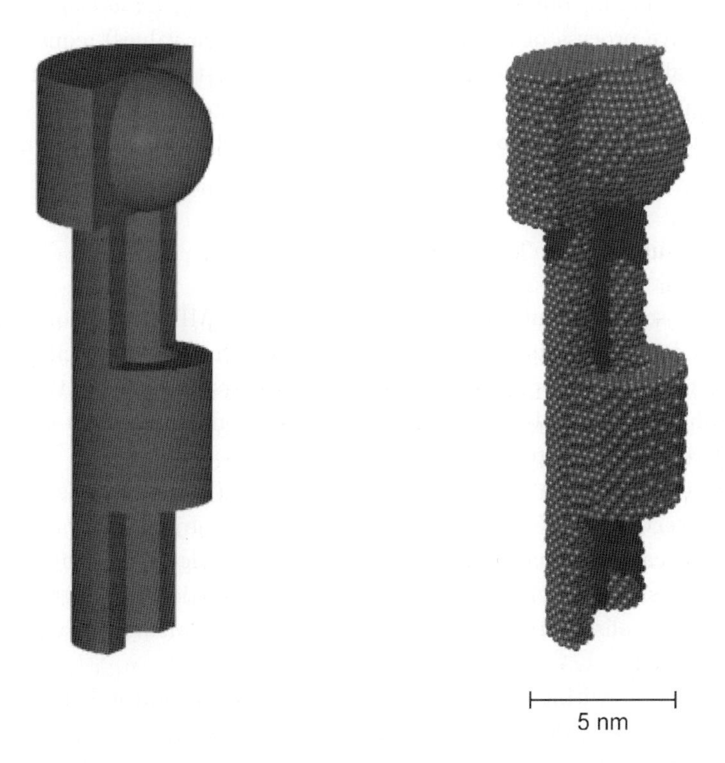

Fig. 3.8 Computer aided nano-design: The left picture shows an object (CAD model) as produced by a common mechanical engineering software. On the right side the according nanostructure is illustrated. It has been created first by filling the hull of the CAD model with (111) oriented aluminium layers, and then by equilibrating the structure at 300 K. The final MD model consists of 16074 aluminium atoms and its height is about 19 nanometer.

Due to the procedure outlined above one could be tempted to the imagination that there is no difference between mechanical engineering and *nano-design* except the different scale units (millimeters or micrometers compared to nanometers), but this would be definitely a fallacy.

While in mechanical engineering edges or surfaces of any shape and continuous calibrations are common, *nano-design* is limited to certain positions of single atoms, i.e. the shape of structures is restricted to certain measures dependent on the material. Moreover, the initial shape of a *nano-model* which may have more or less pronounced edges, can change considerably during the equilibration phase (as well as later on during the MD calculation) and, therefore, the layout of the model often has to be re-designed until it meets the given design criteria.

In conclusion, building MD models or performing *nano-engineering* is an iterative process of *nano-design* (set-up of initial values), equilibration or MD calculation and re-design, and there is almost nothing common with mechanical engineering. More details are illustrated especially in connection with functional nanostructures and nano-machines in Secs. 5.1 and 5.2.

3.2 Visualization Techniques

After the set-up of suitable models (*nano-design*), MD calculations usually produce huge amounts of data, i.e. both, position as well as velocity components are available for thousands of atoms and for thousands of time steps. Though one can apply numerical analysis methods on the MD data, it is far more instructive to visualize the particles.

During the design and construction phase of MD models, it is helpful to get an overview of the boundaries and the different object positions therein. This task can be managed by a more or less simple algorithm that plots the particles as colored circles and transforms the positions orthographically. For more "realistic" representations *rendering* (*ray tracing*) software is the better choice.

Therefore, within the present work a script generator has been implemented that translates the particle positions into a virtual three-dimensional scene description, compatible to the free available ray tracing program *POV–Ray* [Young, 1997; Young, Wells, 1994], according to user specifications. With this, it is possible to define the positions and directions of the camera and light sources to create pictures that convey three-dimensional impressions.

As is well known, single atoms show neither definite surfaces nor colors, their shape, however, might be considered as a cloud resulting from the probability density of the electron positions (orbitals). Since MD treats the atoms as mass points with certain positions and velocities, but without further information about the electrons, it is sufficient to depict the particles in form of colored spheres. It makes sense to adjust the size of the spheres to the "outermost closed electron shell" of the atoms, i.e. to take the value of the atomic or ionic radius (as illustrated in Fig. 3.8 for example), but sometimes it may be more instructive to use smaller spheres and therefore let the structures appear to be porous and transparent. Further, dullness, reflection and transparency of the surfaces of the spheres can be altered — beside colors and a lot of other parameters — by the rendering software. With this, each user has the freedom to realize his subjective imaginations of the atomic representation and to adjust it to his own needs.

The colors may be set either according to distinguish different structure components such as layers, parts of nano-machines, etc. or dependent on the particle velocity to indicate temperature distributions or to visualize the momentum propagation. The latter may be helpful, for example, to analyze impact processes (see Fig. 3.9).

Other important features of the script generator are the ability to perform user-defined cross sections and the selection of certain layers or parts of the simulation space for the representation. This is especially useful to study surface or internal processes which are obscured by the surrounding particles. An example is given in Fig. 3.9.

Applying the rendering technique delivers snapshots of selected MD calculation steps as a result. A further improvement of the visualization is to combine sequences of such snapshots to a movie which can be done easily with the help of numerous free available software tools. In the frame of this work *DTA* [Mason, 1993; Mason, Enzmann, 1993], for example, has been proven as a small, fast and reliable program. The resulting MD movies can be watched with the *AAPlayer* [Autodesk, 1992] that has a lot of additional useful features.

It is clear, but it has to be emphasized: there is nothing that is better suited to understand the huge amount of MD data than a movie. Translations, wave propagation, oscillations, ordering and disordering processes, scattering, generation of dislocations and other defects as well as structural changes or anomalies can be acquired rather by view than by numerical analysis.

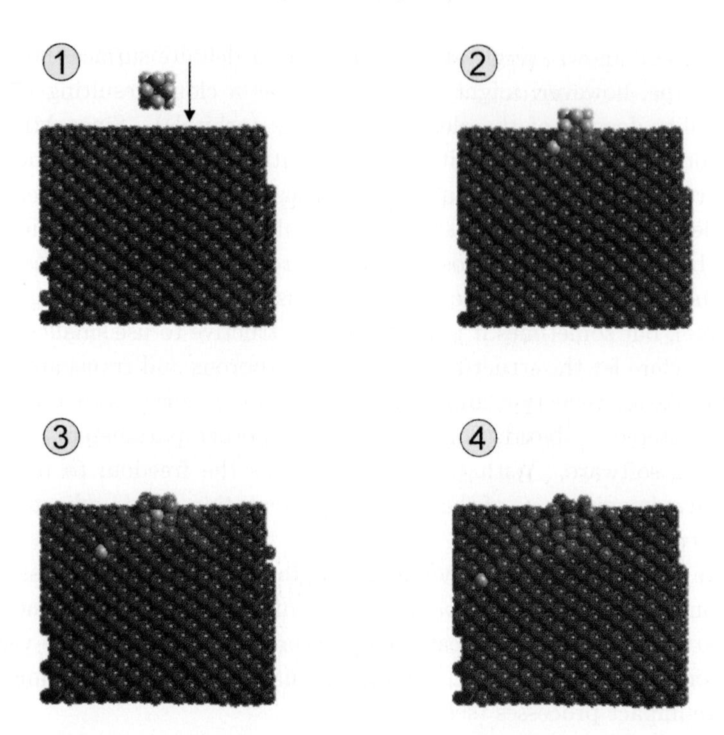

Fig. 3.9 Cluster impact on an aluminium (001) surface: A cross section of the simulation space (periodic boundary conditions) is illustrated. The atoms are colored according to their velocity (from high to low speed: yellow, orange, red, magenta, blue), where yellow corresponds to 20 km/s and blue to room temperature (0.5 km/s). It can be seen quite clearly, that after the penetration of the surface a supersonic shock wave is triggered off which propagates preferably along the $(0 \pm 1 \pm 1)$ directions.

The ultimate visualization technique, however, is to produce pictures that convey a real spatial feeling like that in three-dimensional cinemas. Most often such techniques are connected with more or less extensive hardware such as shutter glasses and special screens or other equipment. But there is a relatively simple method that works with color glasses, e.g. red for the left and blue for the right eye. Watching stereoscopic pictures or *anaglyphs* through such glasses evokes the impression of spatial perception. The production of *anaglyphs* is based on pictures taken from two different positions like, for example, one from the left and one from the right eye. Such pictures can be easily produced from the MD data with the above mentioned rendering software, while the merging procedure into stereoscopic pictures is performed by an *anaglyph generator* (e.g. [SOFTreat, 2001]).

Finally a complete sequence of *anaglyphs* may then be combined to a real three-dimensional movie.

3.3 Solution of the Equations of Motion

Numerical integration methods to solve sets of differential equations can be found in almost any general textbook on applied mathematics and in the more specialized literature on computer simulation or even MD methods (a very small selection is [Allen, Tildesley, 1990; Dahlquist, Björck, 1974; Lapidus, Seinfeld, 1971; Hofer, Lunderstädt, 1975; Hamming, 1973; Gear, 1971; Ralston, Wilf, 1967; Koonin, Meredith, 1990; Haug, 1991; Berendsen, van Gunsteren, 1986; Beeler, Kulcinski, 1972]). All methods are based on finite differences, i.e. the equations are solved step by step in time, where most often the step size $\triangle t$ is taken to be constant.

Now, out of the large class of integration methods we have to select the best fitting algorithm for MD calculations according to the four basic criteria of numerical computation: effectiveness, efficiency, accurateness and stability. But how important are these points for molecular dynamics?

An effective algorithm proves to be fast and requires little memory. In connection with large scale MD studies memory usage may be a limit for the model size depending on the available hardware. Since force computation is by far the most time-consuming part of MD calculations (see following section), the raw speed of an integration algorithm is not an important argument.

But efficiency in the case of MD — defined by the applicable step size $\triangle t$ and by the number of force calculations necessary per time step — takes a key position. First, this means that any integration method that involves more than one force evaluation per step should be excluded. This rules out, for example, all the common *Runge–Kutta* variants. Then, the remaining algorithms should be evaluated according to the size of time steps they are able to employ. In this way, an intended simulation period can be covered in a modest number of integration steps. This in turn leads to acceptable amounts of computation time.

On the other hand, accuracy and stability of a simulation algorithm are measured by its local and global truncation errors, and clearly, the larger $\triangle t$, the less accurate will the MD results follow the classical trajectory (see Fig. 3.10).

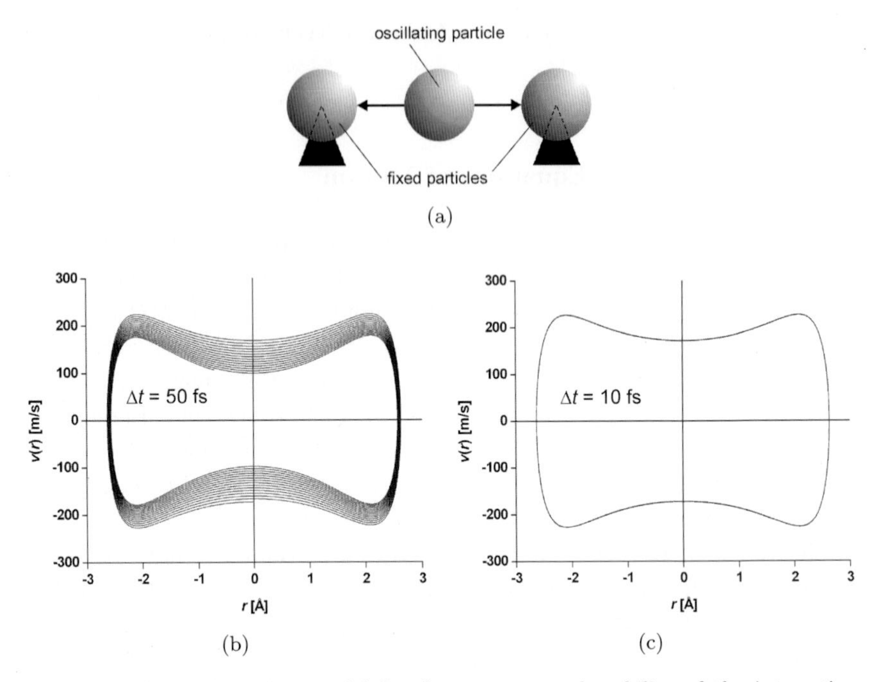

Fig. 3.10 A simple testing model for the accuracy and stability of the integration algorithm. (a): A particle can oscillate between two fixed atoms with respect to the applied potential. The trajectories in the case of krypton are plotted for an initial velocity of 173 m/s (according to 100 K). (b): If the step size $\triangle t$ is chosen to be too large, the computer-generated phase space trajectory (plotted for 2000 steps) diverges from the exact classical one and the total energy is not conserved. (c): With smaller time steps there is no difference between the simulated and classical trajectory, even though here the simulation period has been increased to 10000 steps. In general, the choice for the step size $\triangle t$ strongly depends on the MD model, especially on the particle mass, temperature and interaction potential.

If MD calculations are performed to generate states sampled from an statistical ensemble (*micro-canonical*, *canonical* or *grand canonical*), one does not need the exact classical trajectories, but great emphasis has to be laid on the performance of the conservation laws. Energy conservation, for example, is degraded as the time step is increased. Further, due to its limited stability no integration method will provide an essentially exact solution for an infinite or at least very long time. Compared to the simulation of celestial objects, MD calculations fortunately most often do not need this. Here, in many cases, exact solutions of the equation set of motion are interesting only for periods comparable with the correlation times which, of course, depend on the system under investigation but which usually are not too long. In this sense, the choice of an appropriate integration algorithm involves a trade-off between efficiency and accuracy or stability.

Taking all the above mentioned criteria into account, just a few integration methods remain, because very simple algorithms like *Euler's* and derivatives [Berendsen, van Gunsteren, 1986; Beeler, Kulcinski, 1972] are far too inaccurate. On the other hand, complex methods such as *Runge–Kutta* versions need numerous force calculations per time step.

3.3.1 *Verlet Algorithms*

As a direct solution of the second order equation (Eq. 3.4), the *Verlet* algorithm [Verlet, 1967] is a widely used method to solve the equations of motion. It uses the current position $\mathbf{r}_n$ and acceleration $\mathbf{a}_n$ as well as the previous position $\mathbf{r}_{n-1}$ of an atom to calculate the position $\mathbf{r}_{n+1}$ for the next time step in the following way:

$$\mathbf{r}_{n+1} = 2\mathbf{r}_n - \mathbf{r}_{n-1} + \triangle t^2 \mathbf{a}_n \ , \tag{3.25}$$

with

$$\mathbf{r}_n = \mathbf{r}(t_n) \ , \quad \mathbf{a}_n = \frac{\mathbf{F}(t_n)}{m} \ , \quad t_n = n \, \triangle t \ , \quad n = 0, 1, 2, \cdots \ , \tag{3.26}$$

where the interaction forces $\mathbf{F}$ have to be computed for each particle according to Eq. 3.8. Since the velocities do not appear directly they may be obtained by applying the *central difference method*:

$$\mathbf{v}_n = \frac{\mathbf{r}_{n+1} - \mathbf{r}_{n-1}}{2\triangle t} \ . \tag{3.27}$$

Modifications have been proposed to improve the numerical imprecision of the basic *Verlet* algorithm [Allen, Tildesley, 1990; Dahlquist, Björck, 1974]. One of these is the so-called half-step *leap-frog* scheme [Berendsen, van Gunsteren, 1986; Hockney, 1970]:

$$\begin{aligned} \mathbf{v}_{n+1/2} &= \mathbf{v}_{n-1/2} + \triangle t \; \mathbf{a}_n \ , \\ \mathbf{r}_{n+1} &= \mathbf{r}_n + \triangle t \; \mathbf{v}_{n+1/2} \ . \end{aligned} \tag{3.28}$$

Here the current velocities have to be calculated as mean from the mid-step values

$$\mathbf{v}_n = \frac{1}{2} \left[\mathbf{v}_{n-1/2} + \mathbf{v}_{n+1/2} \right] \ . \tag{3.29}$$

Though at half-steps only, in the *leap-frog* method the velocities appear explicitly. This is necessary in order to perform isothermal equilibration or to sample from *canonical* ensembles which is not possible with the basic form of the *Verlet* algorithm.

Another derivative, the *velocity Verlet algorithm* [Swope *et. al*, 1982], works without mid-steps at the cost of additional storage for $\mathbf{a}_n$:

$$\begin{aligned} \mathbf{r}_{n+1} &= \mathbf{r}_n + \triangle t\ \mathbf{v}_n + \tfrac{1}{2}\triangle t^2 \mathbf{a}_n\ , \\ \mathbf{v}_{n+1} &= \mathbf{v}_n + \tfrac{1}{2}\triangle t\left[\mathbf{a}_n + \mathbf{a}_{n+1}\right]\ . \end{aligned} \tag{3.30}$$

There are still further derivatives [Allen, Tildesley, 1990; Berendsen, van Gunsteren, 1986; Beeman, 1976], but basically all *Verlet* methods produce the same global error and generate identical position trajectories. So there seems to be no need to implement a more complex *Verlet* algorithm than is given by Eq. 3.30.

3.3.2 *Nordsieck/Gear Predictor-Corrector*

Nordsieck [Nordsieck, 1962] and *Gear* [Gear, 1971; Gear, 1966] developed an integration scheme on the basis of *Taylor* expansions of the positions, velocities, accelerations and further derivatives:

$$\begin{aligned} \mathbf{r}(t+\triangle t) &= \mathbf{r}(t) + \triangle t\ \mathbf{v}(t) + \tfrac{1}{2}\triangle t^2 \mathbf{a}(t) + \tfrac{1}{6}\triangle t^3 \mathbf{q}_3(t) \\ &+ \tfrac{1}{24}\triangle t^4 \mathbf{q}_4(t) + \tfrac{1}{120}\triangle t^5 \mathbf{q}_5(t) + \cdots + \tfrac{1}{k!}\triangle t^k \mathbf{q}_k(t) + \cdots\ , \end{aligned} \tag{3.31}$$

$$\begin{aligned} \mathbf{v}(t+\triangle t) &= \mathbf{v}(t) + \triangle t\ \mathbf{a}(t) + \tfrac{1}{2}\triangle t^2 \mathbf{q}_3(t) + \tfrac{1}{6}\triangle t^3 \mathbf{q}_4(t) \\ &+ \tfrac{1}{24}\triangle t^4 \mathbf{q}_5(t) + \cdots + \tfrac{1}{(k-1)!}\triangle t^{k-1} \mathbf{q}_k(t) + \cdots\ , \end{aligned} \tag{3.32}$$

$$\begin{aligned} \mathbf{a}(t+\triangle t) &= \mathbf{a}(t) + \triangle t\ \mathbf{q}_3(t) + \tfrac{1}{2}\triangle t^2 \mathbf{q}_4(t) + \tfrac{1}{6}\triangle t^3 \mathbf{q}_5(t) \\ &+ \cdots + \tfrac{1}{(k-2)!}\triangle t^{k-2} \mathbf{q}_k(t) + \cdots\ , \end{aligned} \tag{3.33}$$

$$\begin{aligned} \mathbf{q}_i(t+\triangle t) &= \mathbf{q}_i(t) + \triangle t\ \mathbf{q}_{i+1}(t) + \tfrac{1}{2}\triangle t^2 \mathbf{q}_{i+2}(t) \\ &+ \cdots + \tfrac{1}{(k-i)!}\triangle t^{k-i} \mathbf{q}_k(t) + \cdots \\ & i = 3, 4, 5, \cdots\ , \end{aligned} \tag{3.34}$$

where

$$\mathbf{q}_k(t) = \frac{\partial^k}{\partial t^k}\mathbf{r}(t)\ . \tag{3.35}$$

Now for the position $\mathbf{r}^{(0)}$ and its *scaled* derivatives $\mathbf{r}^{(k)}$ with

$$\begin{gathered}\mathbf{r}^{(0)} = \mathbf{r},\ \ \mathbf{r}^{(1)} = \triangle t\ \mathbf{v},\ \ \mathbf{r}^{(2)} = \tfrac{1}{2}\triangle t^2\mathbf{a},\\ \mathbf{r}^{(3)} = \tfrac{1}{6}\triangle t^3\tfrac{\partial^3\mathbf{r}}{\partial t^3},\ \ \mathbf{r}^{(k)} = \tfrac{1}{k!}\triangle t^k\mathbf{q}\end{gathered} \tag{3.36}$$

a simple *Taylor* series predictor becomes

$$\begin{bmatrix}\tilde{\mathbf{r}}^{(0)}_{n+1}\\ \tilde{\mathbf{r}}^{(1)}_{n+1}\\ \tilde{\mathbf{r}}^{(2)}_{n+1}\\ \vdots\end{bmatrix} = \mathbf{P}\begin{bmatrix}\mathbf{r}^{(0)}_{n}\\ \mathbf{r}^{(1)}_{n}\\ \mathbf{r}^{(2)}_{n}\\ \vdots\end{bmatrix}, \tag{3.37}$$

where $\mathbf{P}$ is the *Pascal* triangle matrix with the binomial coefficients in its columns:

$$\mathbf{P} = \begin{bmatrix}1 & 1 & 1 & 1 & 1 & 1 & \cdots\\ 0 & 1 & 2 & 3 & 4 & 5 & \cdots\\ 0 & 0 & 1 & 3 & 6 & 10 & \cdots\\ 0 & 0 & 0 & 1 & 4 & 10 & \cdots\\ 0 & 0 & 0 & 0 & 1 & 5 & \cdots\\ 0 & 0 & 0 & 0 & 0 & 1 & \cdots\\ \vdots & \vdots & \vdots & \vdots & \vdots & \vdots & \ddots\end{bmatrix}. \tag{3.38}$$

Due to the missing introduction of the equations of motion, the predictor generates not the exact values for the position and its derivatives. But now, with help of the predicted position $\tilde{\mathbf{r}}^{(0)}_{n+1}$ the forces of the time step $n+1$ can be calculated, and hence the correct accelerations $\mathbf{a}_{n+1}$. The comparison with the predicted (scaled) accelerations $\tilde{\mathbf{r}}^{(2)}_{n+1}$ from Eq. 3.37 gives a measure for the error of the predictor step:

$$\vec{\varepsilon}_{n+1} = \frac{1}{2}\triangle t^2\mathbf{a}_{n+1} - \tilde{\mathbf{r}}^{(2)}_{n+1}\ . \tag{3.39}$$

Then this error is used to improve the predicted values in a corrector step which reads:

$$\begin{bmatrix} \mathbf{r}_{n+1}^{(0)} \\ \mathbf{r}_{n+1}^{(1)} \\ \mathbf{r}_{n+1}^{(2)} \\ \vdots \end{bmatrix} = \begin{bmatrix} \tilde{\mathbf{r}}_{n+1}^{(0)} \\ \tilde{\mathbf{r}}_{n+1}^{(1)} \\ \tilde{\mathbf{r}}_{n+1}^{(2)} \\ \vdots \end{bmatrix} + \begin{bmatrix} c_0 \\ c_1 \\ c_2 \\ \vdots \end{bmatrix} \vec{\varepsilon}_{n+1} \, . \tag{3.40}$$

The values for the corrector vector have to be chosen according to the expansion i of the *Taylor* series in Eqs. 3.31–3.34. Usually the *Nordsieck/Gear* algorithm works with $i = 3, \cdots, 8$ values for which the corrector vectors can be found, e.g. in [Allen, Tildesley, 1990; Berendsen, van Gunsteren, 1986; Beeler, Kulcinski, 1972; van Gunsteren, 1977]. A compilation is given in Tables 3.1 and 3.2.

Further representations of this scheme, where the higher derivatives are replaced by the accelerations of prior steps, can be found in [Berendsen, van Gunsteren, 1986; van Gunsteren, 1977].

Table 3.1 Nordsieck/Gear corrector vectors [Allen, Tildesley, 1990; Berendsen, van Gunsteren, 1986].

i	c_0	c_1	c_2	c_3
3	0	1	1	
4	1/6	5/6	1	1/3
5	19/120	3/4	1	1/2
6	3/20	251/360	1	11/18
7	863/6048	665/1008	1	25/36
8	275/2016	19087/30240	1	137/180

Table 3.2 Nordsieck/Gear corrector vectors (*continued*).

i	c_4	c_5	c_6	c_7
5	1/12			
6	1/6	1/60		
7	35/144	1/24	1/360	
8	5/16	17/240	1/120	1/2520

Basically, within predictor-corrector methods the corrector step may be repeated to refine the results. Fortunately, due to its accuracy this is not necessary for the *Nordsieck/Gear* algorithm. In this way, as already mentioned, the huge effort of an additional force computation is avoided.

To gain the real velocities $\mathbf{v}_{n+1}$ — for example to calculate the temperature — the according part of Eq. 3.36 has to be reversed:

$$\mathbf{v}_{n+1} = \frac{\mathbf{r}_{n+1}^{(1)}}{\triangle t} . \tag{3.41}$$

Further, it is worth to mention that the *Nordsieck/Gear* scheme is a self-starting algorithm, i.e. it is sufficient to give the initial positions and velocities and set the higher derivatives to zero at the beginning.

3.3.3 *Assessment of the Integration Algorithms*

The most favorable algorithm with respect to the required memory storage is the *leap-frog scheme* with three storage vectors, followed by the *velocity Verlet* with four and the *k-value Nordsieck/Gear* algorithm with $k+1$ vectors. This may become important for studies of large-scale MD models.

In [Berendsen, van Gunsteren, 1986] a comparison of accuracy, energy drift and mean square fluctuations dependent on the step size $\triangle t$ for the case of the harmonic oscillator potential has been illustrated. Though a simple model, the results in principle show the behavior of the different integration algorithms, where for all step sizes the *Nordsieck/Gear* algorithm clearly has shown the highest accuracy with respect to deviations from the exact trajectory.

Considering the conservation or fluctuations of the total energy leads to mixed results. Especially with real models, i.e. simulations of complex systems, there is an upper limit for the step size above which the *Verlet* methods are superior to the *Nordsieck/Gear* algorithms [Allen, Tildesley, 1990]. Further, the latter reach an optimum with respect to energy fluctuations for $k = 6$ or 7. Actually, there is just a very small difference between $k = 6$ and 7.

While speed of the integration schemes generally plays a minor role, this leads to two conclusions:

(1) If accuracy and long periods are not important for the simulations, the *Verlet* algorithms have to be preferred.

(2) For high-accuracy problems or long-time simulations the 6-*value Nordsieck/Gear predictor-corrector* yields better results, though at the cost of decreased step sizes.

Therefore, the MD calculations of the current work most often have been performed using the 6-*value Nordsieck/Gear algorithm.*

3.3.4 *Other Methods*

Though the described methods (including their derivatives) are the most frequently used and appropriate ones, we would like to mention at least *Rahman*, who — to our knowledge — has performed the earliest realistic MD study with continuous potentials for the example of liquid argon [Rahman, 1964]. In this work he utilized a simple predictor-corrector method that has the disadvantage, in that it needs several corrector passes to provide accurate solutions of the equations of motion. In his later works, he switched over to *Nordsieck/Gear* algorithms [Rahman, Stillinger, 1971].

Still, there are several other methods that claim an improved behavior with respect to the one or other characteristic, for which [Fincham, Heyes, 1982; Heyes, Singer, 1982] include some examples.

3.3.5 *Normalized Quantities*

When performing numerical calculations usually the use of normalized quantities is handier compared to standard unit systems like, for example, the *cgs* or *mks* system [Physikalisch Technische Bundesanstalt, 1985]. A normalized quantity $\hat{x}$ is derived from the real value x by a scaling factor x_0: $\hat{x} = x/x_0$. There are a lot of scaling possibilities (e.g. the application of atomic units), but due to the special characteristics of MD models we have chosen the unit system as given in Table 3.3.

Table 3.3 Units and scaling factors for normalized quantities.

Mass m_0	Time t_0	Distance r_0	Velocity v_0	Force F_0	Energy E_0
10^{-25} kg	10^{-14} s	10^{-10} m	10^4 m/s	10^{-7} N	10^{-17} J
60.22 a.m.	10 fs	1 Å	10 Å/ps	0.1 μN	62.41 eV

3.4 Efficient Force Field Computation

Without question, the most time-consuming part of MD simulations is the computation of the particle interactions. Obviously, when considering Eq. 3.8, the reason for this are the numerous summations over all two-body forces f_{ij} that are necessary to yield the resulting force $\mathbf{F}_i$ for the ith particle. For a system of N atoms the number of single summation steps is $N(N-1)$. Using the fact, $f_{ij} = f_{ji}$ halves this number (*Newton's third law*).

Software algorithms handle the force computation with the help of two nested loops, where the outer loop counts over all atoms while the inner one considers the interacting particles only. In this way the necessary pairs of indices ij are available to calculate the particle distances, pair potentials, resulting forces, directions and vector components.

Thus, designing an efficient — and therefore fast — algorithm for the computation of the N-body force field in any case means to reduce the instructions of the inner loop to a minimum. Since that mainly concerns the derivation of particle acceleration components from a given pair potential, the subject will be discussed next.

Beside this, there are some sophisticated techniques that reduce the number of passes through the inner loop significantly. Basing on the finite range of the interactions, these methods are the key to large-scale MD simulations.

3.4.1 *Force Derivation*

If a pair potential is given analytically, then the according interaction force can be easily derived analytically (Eq. 3.6). But, however simple the resulting expression may be (in the case of a *Barker* or *Schommers* potential it is rather complex), its application leads not exactly to a reduction of the evaluations necessary within the inner loop.

A common procedure within numerical algorithms is to trade memory for speed. Here this means that the computation speed can be increased by the storage of the pair interaction force as a function of particle distance:

$$f_k = f(r_k) = -\left.\frac{\partial v(r)}{\partial r}\right|_{r=r_k} , \quad r_k = r_0 + k\,\triangle r \; , \quad k = 0, 1, 2, \cdots \; , \qquad (3.42)$$

where r_0 is the minimum distance and $\triangle r$ is the table spacing.

Such a force table f_k is prepared in advance, either according to an analytical expression or numerically, which can be used further during the

MD calculation. Usually a spacing of $\triangle r = 10^{-3}$ Å is sufficient to avoid an interpolation of values lying between those of the table. A numerical derivation of the force table from the potential v may be performed by the simple *central difference method*:

$$f_k = \frac{v(r_{k-1}) - v(r_{k+1})}{2\triangle r} \ , \quad r_k = r_0 + k \triangle r \ , \quad k = 0, 1, 2, \cdots \ . \tag{3.43}$$

There is still room for further fine tuning. An example is to avoid the square root, necessary to obtain the particle distance, which is relatively time consuming on several computers. This can be done by scaling the force with $r^{1/2}$ and storing the table as a function of r^2. Then the square root in the Eq. 3.8 vanishes. Another, if slighter improvement is to work with an acceleration table by scaling the force table with the inverse particle mass. This makes the according division operations obsolete within the integration algorithm.

3.4.2 *List Method*

However, the most sophisticated algorithms with respect to a significant reduction of the passes through the inner force computation loop are based on the artificial restriction of the interaction range. Therefore, the pair potential or the force table has to be multiplied with a cut-off function as, for example, that given in Eq. 3.24 in connection with periodic boundary conditions. Here in addition, cutting-off is necessary to avoid numerical instabilities as well as to increase the energy conservation behavior. Alternative approaches are given in [Allen, Tildesley, 1990; Streett *et. al*, 1978; Stoddard *et. al*, 1973; Powles *et. al*, 1982].

Now, using potentials with a maximum range of r_c during the $(N-1)$ passes through the inner loop, force computations can be restricted to those particles that are within the interaction range, i.e. forces are calculated only, if the particle distance is smaller than r_c. Though this doesn't reduce the number of loop passes but further needs a condition command, the total number of executed instructions is much smaller depending on the cut-off radius r_c.

The *list method* [Allen, Tildesley, 1990; Verlet, 1967] goes one step further, again, trading memory for speed. Here, for each particle a list is stored that contains its neighbors within a sphere of radius $r_l > r_c$. That is, the inner loop treats just these particles which are about within interaction range. Due to the particle motion the *neighbor lists* have to be updated from time to time, where the interval depends on the MD model as well as on the list sphere radius r_l. But usually, after about each 20 calculation steps an update should be performed. Another possibility is to make the update interval dependent on the particle displacements [Fincham, Ralston, 1981; Thompson, 1983].

However, compared to the standard procedure the *list method* reduces the number of force computations n_F significantly (see Fig. 3.13):

$$n_F \approx \frac{N^2}{2\triangle s} + N_l N , \tag{3.44}$$

where $\triangle s$ denotes the number of calculation steps of the update interval and N_l is the average number of particles within the *neighbor list*.

3.4.3 *Cell Algorithms*

As the system size increases, the memory usage of the *list method* may become too large, and the update procedure — depending on N^2 — needs too much time.

In this case *cell algorithms* are even more efficient. Here, the simulation space first is divided equally into cubical cells (see Fig. 3.11). Then the particles are assigned to the single cells according to their positions, where the assignments are stored in a list for each cell. Here, it is important to

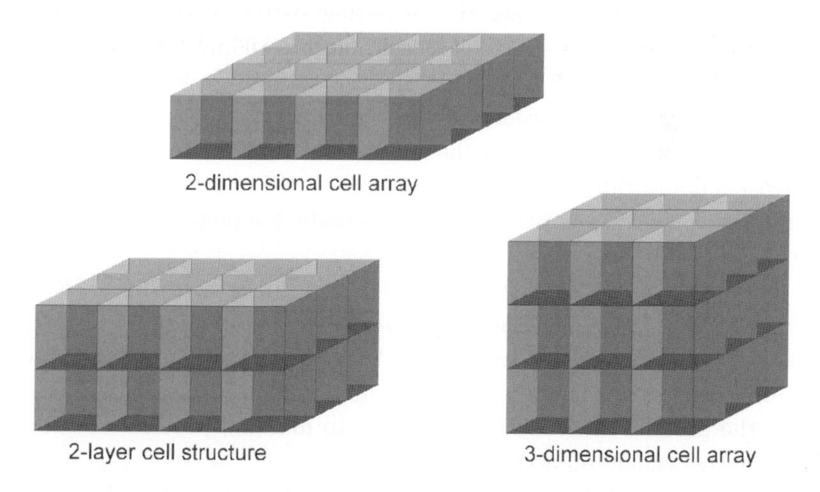

Fig. 3.11 The cell algorithm bases on the division of the simulation space into small cells with side lengths greater than the cut-off distance r_c. Different examples for possible arrangements are illustrated.

choose the side lengths of the cells slightly greater than the cut-off distance r_c of the interaction potential.

With such a cell arrangement the force computation has to be performed in two steps. First, the interactions of particles within the same cell are considered, followed by the force computation for atoms located in neighboring cells, where the number of direct cell neighbors n_n depends on the cell arrangement and on the location (boundary cells have less neighbors, see Fig. 3.11). However, neglecting cells located at the boundary, for (large) two-dimensional arrays $n_n = 8$, for two-layer structures $n_n = 17$, and for three-dimensional arrangements $n_n = 26$. Therefore, using the cell algorithm the total number of force computations n_F is given by:

$$n_F \approx (n_n + 1) N_c N \ , \tag{3.45}$$

where N_c is the average number of particles per cell.

The assignment of the particles to the according cell lists is easy to implement, needs less computation time and, therefore, may be performed after each MD calculation step.

3.4.4 *SPSM Procedure*

When it comes to periodic boundary conditions and/or parallel computing, the cell algorithms are superior to all the other methods.

The virtual mirroring into the according spatial directions of all particles, as described in a previous section, now is reduced just to particles within boundary cells. That is, for the interaction of a boundary cell not only the direct neighbor cells are taken into account, but all its counterparts on the opposing sides of the simulation box. In this way, in a two-dimensional and three-dimensional array each cell has 8 and 26 neighbors, respectively, i.e. cell algorithms provide periodic boundary conditions without further effort.

One of the most sophisticated cell algorithm is the *SPSM* (*Scalable Parallel Short-Range Molecular Dynamics*) *method* [Lomdahl *et. al*, 1993; Beazley, Lomdahl, 1994]. It is tailored for the use of parallel computers since its code scales linearly with the number of processors and particles, while the parallel efficiency (with respect to minimal processor communication rates) ranges on a very high level.

In contrast to the standard *cell algorithm*, the application of the *SPSM method* bases on a division of the simulation space into rectangular subspaces, prior to the further subdivision into cells. The subspaces are

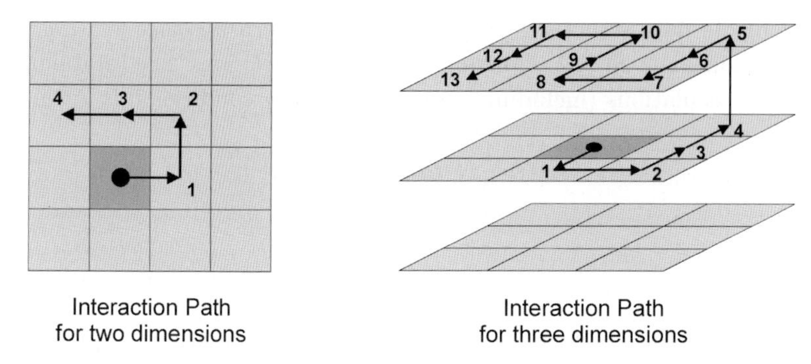

Fig. 3.12 Applying Newton's third law on the computation of particle interactions between neighboring cells reduces the number of force calculations. Instead of considering all, it is sufficient to include just one half of the direct cell neighbors which can be determined by following specific interaction paths. The most obvious choices for such interaction paths are illustrated (see [Lomdahl *et. al*, 1993; Beazley, Lomdahl, 1994]), others are possible, too.

assigned to the processing nodes of a parallel machine, where the arrangement of the subspaces plays no major role (a subspace may differ in its length, height and depth).

The force computation is basically the same like that described above, but for a further reduction of runs through the inner loop, *Newton's third law* (in this context $f_{ij} = f_{ji}$) is applied to the interactions of neighboring cells. In the case of a cell structured calculation space this means, that only *one half* of the cell neighbors have to be considered. These neighbors are selected by following a specific "interaction path" (see Fig 3.12). The interactions resulting from the other, missing half of neighbors are considered when the path is followed starting from such a neighbor.

With that, the *SPSM* scheme works as follows. First all interactions of particles within the same cell are computed. Then, forces between particles in neighboring cells are calculated following the interaction path. In this way, the accelerations are accumulated by the original cell and each "visited" cell. In order to calculate all forces (including those resulting from neighbors that are not on the interaction path) this procedure is carried out on all cells and simultaneously on all processor nodes.

Whenever the interaction path crosses a processor boundary, *message passing* is used to communicate particle data. Here it is the main advantage of the *SPSM* scheme that it can be translated into a very efficient code, where each processor node simultaneously manages its own cells as well as those received from its neighbor nodes. Moreover, the use of appropriate data structures combined with modern *message passing* techniques makes the *SPSM* algorithm without competition. The total number of force calculations (including periodic boundary conditions) is given by

$$n_F \approx \frac{1}{2} N \left[(n_n + 1) N_c - 1\right] . \tag{3.46}$$

3.4.5 *Discussion*

In Fig. 3.13, the approximate number of force calculations is plotted for each method as a function of the particle number. As can be seen quite clearly, the standard "brute force" method should never be used. For MD models consisting of particles in the range of up to several ten-thousands the

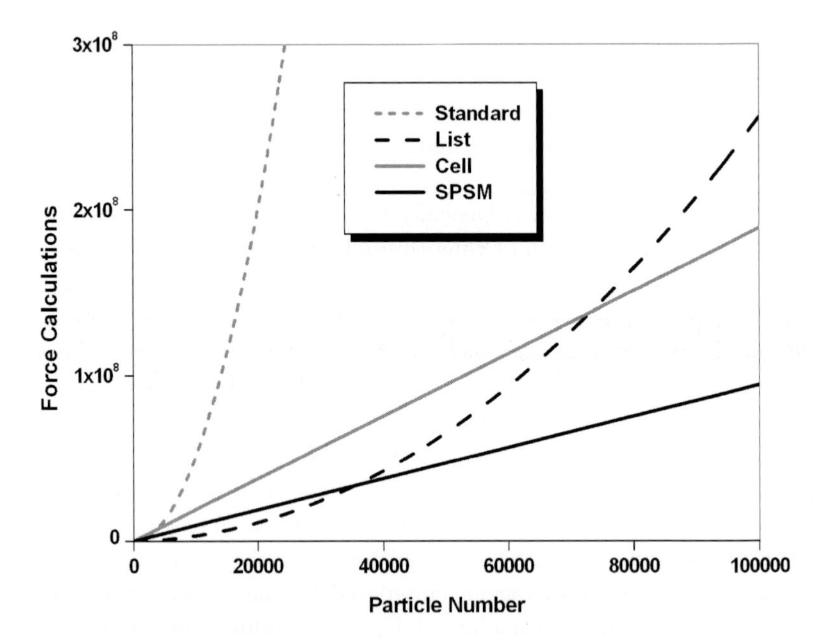

Fig. 3.13 Comparison of different force computation methods. The approximate numbers of force calculations for a system of N particles are illustrated as functions of N. The parameters of the list method have been chosen to be $\triangle s = 10$ (update interval) and $N_l = 60$ (average number per neighbor list). For the cell algorithm and SPSM method the average particle number per cell N_c is 70 and the next neighbor number n_n is 26 according to a three-dimensional cell arrangement. All parameters represent approximately a bulk model for aluminium. The overall timing of real algorithm implementations may differ from the plot, but basically the dependence on the particle number is the same.

list method performs best. Above this range — it depends strongly on the MD model, on the implementation of the algorithms, and on the computer hardware — the *SPSM* procedure is unbeatable. Though the standard cell algorithm too performs well for large-scale simulations, it is always worth to spend the additional effort, necessary to implement the *SPSM* scheme.

3.5 Implementation

For the MD studies, performed in the frame of this monograph, it was necessary to develop a rather complex software system (see Fig. 3.14). The basic concept — to be as flexible as possible — consists of the integration of the modeling and visualization modules in a common PC program, while the MD calculation and analysis algorithms have been separated for the use on different computer platforms.

The PC modules (nano-engineering, potential generator, visualization)

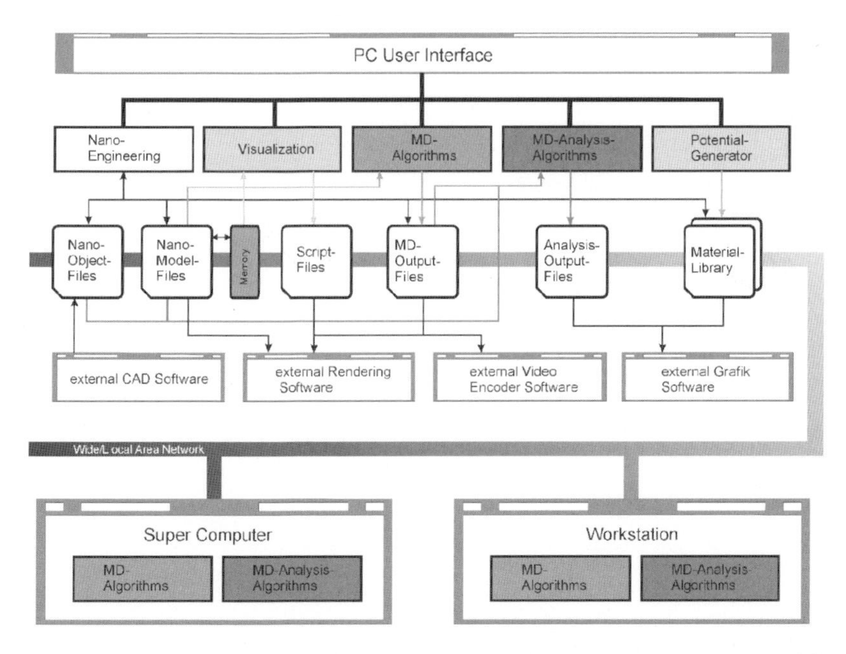

Fig. 3.14 Schematic drawing of the developed MD software system: The essential features have been integrated in a PC based software. The algorithms for MD calculation and analysis are implemented on super computers and on workstations. A common file system is the interface for distributed MD computing via LAN/WAN or for the graphical processing with the various external software systems.

have been developed with the Microsoft Visual Studio, and a window based user interface allows for easy set-ups of different MD models and for a graphical processing of the MD data.

The MD and analysis algorithm developments have been performed using the FORTRAN90 programming language with the additional MPI library for parallel computing. Therefore, the same code could be used for compilations on various platforms. Several MD algorithms (*Verlet*, *Nordsieck/Gear* and *Rahman* integration methods combined with *list*, *cell* and *SPSM* force calculation schemes) have been implemented on PC and UNIX workstations, on a SNI VPP300-32 vector parallel machine and on an IBM SP2-256 massive parallel computer.

Since all interfaces have been designed in the form of a common file system, the communication and data transfer can be handled either via the *Local* or *Wide Area Network* (*LAN/WAN*) to perform distributed computations.

Chapter 4
Characterization of Nano-Systems

Just a few years ago, materials research had mainly to do with macroscopic systems. The situation changed fundamentally with the rise of nanotechnology. Now properties which could have been well defined and clearly fixed for macroscopic (bulk) systems no longer can be assigned to nanostructures — at least not in the same simple way.

In this section, we would like to discuss this problematic nature connected to the characterization of nanosystems. First, the difficulties are outlined at the example of the thermal stability of nanostructures, then we go further into detail considering other basic material properties. In this context, it will become apparent that the standard model of solid-state physics fails at the nanometer level, that is, characterization of nanostructures has to be performed by other means. The alternatives, arising from the analysis of MD calculations, are compiled in the final section, and wherever possible, we illustrate the topics with exemplary MD studies.

4.1 Thermal Stability

Melting temperature and *thermal stability* of nanosystems are examples for such properties which may change extremely when we go from the macroscopic to the microscopic realm, i.e. when we consider nanometer-scale properties. The melting temperature of macroscopic systems is well defined and the structures of these systems are thermally stable up to the melting point. Of course, phase transitions may occur, but the various phases define stable configurations as well.

In contrast to macroscopic systems, the melting temperature of nanosystems depends on the particle number and is also a function of the shape of the system. As will be demonstrated, the melting temperature of certain nanosystems is not defined and not clearly fixed, respectively. Especially in connection with materials research, this is a completely new situation.

Let us demonstrate this point by means of an example: the melting temperature of aluminium (Al) is 933 K and the microscopic structure and outer shape of macroscopic Al-systems are stable up to 933 K. Typically, this is not the case for Al-systems of nanometer-size.

We performed MD calculations, as has been shown in the prior sections, on the basis of the realistic *Schommers* pair potential. Here, the following results are of particular interest [Rieth *et. al*, 2000; Rieth *et. al*, 2001]:

The *thermal stability* of structures on the surface — as, for example, the three-dimensional object in Fig. 4.1 — in general will only be weakly correlated with the melting temperature. The structures are usually unstable below the melting point and may even dissolve.

In Fig. 4.1, a three-dimensional "F"-shaped structure consisting of aluminium atoms is shown. Though the melting point of Al is 933 K, it can be clearly seen from Part (4) of the figure that this configuration is structurally disturbed already at $T = 270\ K$, which is significantly below the melting temperature.

The system develops as follows: Part (1) of Fig. 4.1 represents the initial configuration. It corresponds to the crystalline structure of aluminium at zero temperature. The time step used in the calculations is $5 \cdot 10^{-15}$ sec; the particle number of the nanosystem is $N = 1660$ (without substrate). After 2000 time steps the system has reached a temperature of 250 K [Part (2) of Fig. 4.1]. After 6000 time steps [Part (3) of Fig. 4.1] the temperature is 270 K, and the system remains at this temperature, but there are still changes in connection with the shape of the system. After 10^4 time steps the outer form remains constant too, as is shown by Part (4) of the figure.

It is typical for the behavior of such nanostructures that a tiny variation of the initial conditions leads to different final shapes. Two further examples are illustrated in Fig. 4.2.

In connection with the studies in Figs. 4.1 and 4.2, the following is essential: the thermal behavior is not entirely determined by the number of particles, but also by the outer form of the nanosystem.

In order to investigate the melting process of the structure given in Figs. 4.1 and 4.2, the temperature has been increased (see Fig. 4.3). It turned out that a melting temperature is not or just hardly definable, because the onset of sublimation is prior to the melting process, that is, the

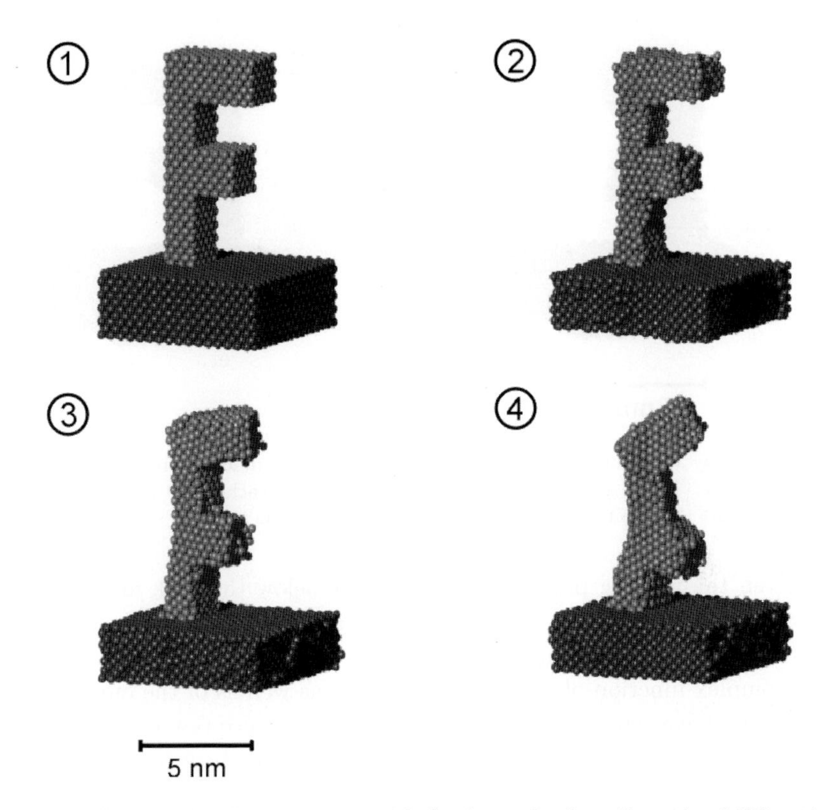

Fig. 4.1 A MD study of a nanosystem with the shape of a three-dimensional "F" resting on a substrate. Both the nanostructure and the substrate consist of aluminium. The "F" is built up with 1660 atoms. Starting from 0 K [Part (1)] the temperature rises continuously up to 270 K [Part (3)], whereas the change of the outer shape still continues. Part (4) shows the point at which the system reaches an equilibrium.

melting process is "overtaken" by the sublimation process and, therefore, melting does not take place.

In conclusion, specific material properties of nanosystems may differ essentially from the corresponding properties of macroscopic systems. This has been demonstrated in connection with the thermal stability and the melting temperature of the nanostructure shown in Fig. 4.1.

The main reason for this tendency is the fact that a great fraction of the particles (atoms, molecules) of such small systems belongs to the surface region. Since the surface particles are less bonded than the particles in the bulk, this leads to relatively strong anharmonicities already at low temperatures, that is, far below the melting temperature.

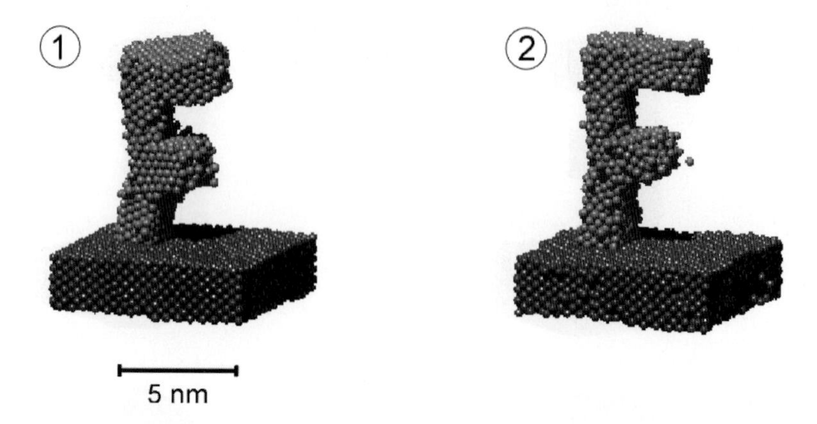

Fig. 4.2 Tiny variations of the initial conditions lead to different final shapes of the nanostructure. In the case of Part (1), the temperature reaches 400 K after 10^4 time steps. In Part (2), the temperature is 500 K after 5000 time steps.

Even the melting process takes place far below the bulk melting temperature or is simply not defined as has been demonstrated in connection with the "F"-shaped nanosystem. The thermal behavior of such systems is a complex function of the particle number as well as of the outer shape. This must have consequences for the theoretical description of the material properties for systems of nanometer-size as we will show in the following section.

4.2 Basic Material Properties

In modern materials research solid state physics became more and more relevant, in particular, the microscopic structure and dynamics. For sufficiently large systems the *standard model* of *solid state physics* is in most cases adequate for the description of material properties. The standard model of solid state physics bases on the *ordered structure* and on the assumption that the vibrational amplitudes of the atoms are sufficiently small, so that it can work within the *harmonic approximation*. In other words, the dynamics is expressed by *phonons*. For example, the specific heat at constant volume is expressed in terms of phonons by the following expression [Allen, de Wette, 1969; Maradudin *et. al*, 1967]:

$$c_V = k_B \sum_{p,q} \frac{\alpha^2 e^\alpha}{(e^\alpha - 1)^2} \tag{4.1}$$

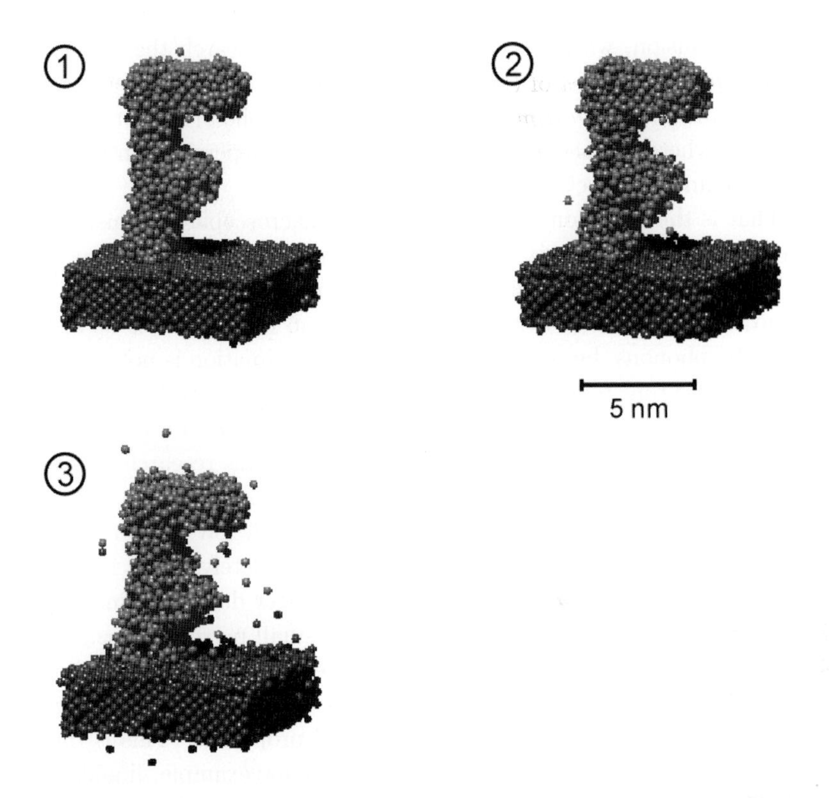

Fig. 4.3 An investigation of the melting process of the nanostructure given in Figs. 4.1 and 4.2: Starting point is the configuration as illustrated in Part (2) of Fig. 4.2. First the temperature has been increased to 600 K. The result after 5000 time steps is shown in Part (1). There is no significant change to recognize. The second attempt consists of an increase to 750 K. The development of the nanostructure after 1000 calculation steps is shown in Part (2). After 4000 time steps the temperature is still at 750 K and the onset of sublimation can be clearly seen [Part (3)]; the melting process is overtaken by the sublimation and a melting temperature for the nanostructure is not definable.

with

$$\alpha = \frac{\hbar\omega_p(q)}{k_B T} , \tag{4.2}$$

where $\omega_p(q)$ are the phonon frequencies, q is the wave vector and p labels the phonon branches.

Of course, electronic properties too are essential. An example is the understanding of superconductivity, where the electron–phonon interaction is important.

In conclusion, within *conventional* materials research the basis of the *microscopic description* of the material properties is the ordered structure and the *harmonic approximation.* Properties of a solid are given in terms of phonons, where the phonon frequencies $\omega_p(q)$ are dependent on the crystal structure and the electronic properties.

That is the situation in connection with macroscopic systems, but this can no longer be the basis for *materials research in nanotechnology.*

As we have seen above, nanosystems do not behave in such a simple way. In most cases, they are neither ordered nor can the dynamics be approximated by phonons, because the harmonic approximation is not applicable. Even the melting temperature is no longer a fixed material property, as we demonstrated in connection with the nanostructure on a substrate in Fig 3.1. Therefore, at the nanometer level materials research gets a new dimension.

Nanosystems are disordered already at relatively low temperatures and the *harmonic approximation* breaks down completely, that is, even at relatively low temperatures *phonons* are no longer definable since the anharmonicities can no longer be considered as small perturbations.

In other words, at the nanometer scale the *standard model* of solid state physics breaks down in many cases and we have to introduce other methods and quantities for the description of material properties. This is the case for any anharmonic and disordered system, as for example, liquids in the bulk. For simplicity, we would like to explain the principal points by means of a bulk liquid.

In the case of liquids the usual crystallography and the phonons are not relevant. The structure has to be described rather by statistical mechanics, that is, in terms of correlation functions, as for example, the pair correlation function $g(r)$ which can be measured (r is the relative distance between two particles). Furthermore, we have to know the interaction potential between the atoms and molecules, respectively. In the case of metals, the interaction is given by the pair potential $v(r)$, where again r is the relative distance between two particles.

In conclusion, instead of the phonons $\omega_p(q)$, correlation functions ($g(r)$ etc.) and the interaction potential $v(r)$ become the relevant quantities: $\omega_p(q) \mapsto g(r), v(r)$.

For example, the isothermal compressibility χ_T is defined by the following equation [March, 1968]:

$$\frac{1}{\chi_T} = -V \left(\frac{\partial p}{\partial V} \right)_T . \tag{4.3}$$

Now we need an expression for the pressure p (V is the volume), that is, we need the *equation of state*, which can be formulated in terms of $g(r)$ and $v(r)$ as follows:

$$p = \rho k_B T - \frac{\rho^2}{6} \int r \frac{\partial v(r)}{\partial r} g(r) d\mathbf{r} . \tag{4.4}$$

If Eq. 4.4 is applied on Eq. 4.3, the isothermal compressibility χ_T can be expressed in terms of $g(r)$ and $v(r)$: $g(r), v(r) \mapsto \chi_T$.

As already mentioned, in the case of nanosystems the anharmonicities are very strong. Moreover, a great fraction of the atoms belongs to the surface region and it turned out that the thermal expansion at the surface is distinctly larger than in the bulk. Therefore, the thermal expansion too becomes a relevant quantity and the theoretical picture should be able to describe it.

The thermal expansion coefficient α_p can be expressed by the isothermal compressibility χ_T and the thermal pressure coefficient γ_V:

$$\alpha_p = \chi_T \, \gamma_V . \tag{4.5}$$

In order to get the thermal expansion coefficient α_p we just have to calculate γ_V, since χ_T is already given by Eq. 4.3:

$$\gamma_V = \left(\frac{\partial p}{\partial T} \right)_V . \tag{4.6}$$

In addition, we need the equation of state for the pressure p which is given for the bulk by Eq. 4.4. In other words, the thermal expansion coefficient α_p too can be expressed in terms of $g(r)$ and $v(r)$: $g(r), v(r) \mapsto \alpha_p$.

Basically, this is the case for all material properties except that in some cases higher order correlation functions are needed.

It should be mentioned that the thermal expansion is zero for the harmonic case, that is, within the phonon picture the thermal expansion cannot be described.

In conclusion, in the case of disordered systems with strong anharmonicities like, for example, liquids in the bulk, phonons are not suitable for an adequate description, but the according expressions have to be formulated in terms of $g(r)$ and $v(r)$.

Since in the case of nanosystems, a great fraction of particles belongs to the surface regions, here the quantities $g(r)$ and $v(r)$ are not only dependent on the relative particle distance r. In addition, the distance of particles from the surface has to be taken into account. Usually there is more than one surface in the vicinity of a particle (see, for example, Figs. 4.1–4.3), so the situation may become rather complex.

Even for systems with one surface only — for example, a semi-infinite liquid — the situation is already complicated. In this case, the surface (defining the x, y-plane) separates the liquid from the vapor phase and the density ρ varies along the z-axis (the co-ordinate perpendicular to the surface), so that we have $\rho = \rho(z)$.

Since p is neither a constant nor independent on the coordinates in the liquid–vapour transition zone, the pressure cannot be described by Eq. 4.4. Here p is an anisotropic quantity [Schommers, 1987], that is, at any point z in the transition zone we have a normal pressure component p_n and a tangential component p_t. Both are different from each other in the transition zone (of course, they are identical in the isotropic bulk phase). With $\mathbf{r} = (x_{12}, y_{12}, z_{12})$, the statistical mechanical expressions for p_n and p_t take the form [Schommers, 1987; Croxton, 1980]

$$p_n = \rho(z)k_BT - \frac{1}{2}\int \rho^2(z)\rho^{(2)}(\mathbf{r}, z)\frac{\partial v(r)}{\partial r}\frac{z_{12}^2}{r}d\mathbf{r}\,, \tag{4.7}$$

$$p_t = \rho(z)k_BT - \frac{1}{2}\int \rho^2(z)\rho^{(2)}(\mathbf{r}, z)\frac{\partial v(r)}{\partial r}\frac{x_{12}^2}{r}d\mathbf{r}\,. \tag{4.8}$$

Here, for simplicity, it is assumed that the pair potential at the surface is the same as in the bulk and instead of the correlation function $g(\mathbf{r}, z)$ in Eqs. 4.7 and 4.8, the two-particle distribution function $\rho^{(2)}$ is used.

With the pressure components p_n and p_t, the surface tension γ may be determined by

$$\gamma(T) = \int_{-\infty}^{\infty} |p_n - p_t(z)| dz\,, \tag{4.9}$$

and using Eqs. 4.7 and 4.8, the statistical mechanical expression becomes

$$\gamma(T) = \frac{1}{2}\int_{-\infty}^{\infty} \rho^{(2)}(\mathbf{r}, z)\frac{\partial v(r)}{\partial r}\frac{x_{12} - z_{12}}{r}\, d\mathbf{r}\, dz\,. \tag{4.10}$$

The surface tension and its temperature dependence is of particular interest in the description of thermodynamic properties [Schommers, 1987].

In the case of nanosystems, like those given in Fig. 4.1, the situation with respect to statistical mechanical descriptions is more complex than that for semi-infinite systems. This is due to the fact that here various surfaces have to be considered which are closely arranged in the form of more or less complicated geometries. Therefore, the statistical mechanical expressions (in analogy to Eqs. 4.7, 4.8 and 4.10) will hardly be relevant in connection with numerical calculations. The more direct way is to perform *molecular dynamics* calculations.

4.3 Wear at the Nanometer Level

Within the frame of such molecular dynamics systems, friction in the macroscopic sense is not defined. At the microscopic molecular dynamics level the forces are formulated as quantities which are dependent on the structural configuration (particle positions) but not on the particle velocities. Therefore, a force which is proportional to the velocity cannot be introduced at the microscopic level and therefore, a friction constant in the macroscopic sense is not definable at the microscopic level in modern materials research; at this level wear is described by specific complex processes. This point may be discussed best by means of an example:

Figure 4.4 shows a molecular dynamics model for a spinning wheel moving towards a thin film. The wheel rotates with about 10^{12} revolutions per second, has a temperature of 300 K and its diameter is approximately 10 nm. When the wheel contacts the surface, friction-effects emerge and according to the magnitude of the vertically applied force the wheel may even be destroyed as is illustrated in Fig. 4.4.

In conclusion, friction at the microscopic level is a complex process and cannot be characterized by one constant only as is the case for macroscopic friction. In the example of Fig. 4.4, wear is dependent on the specific structure of the surface and additionally on the shape and motion of the wheel.

4.4 Mean Values and Correlation Functions

As has been shown, materials research at the nanometer level must not be restricted to conventional solid state physics which is based on an ordered

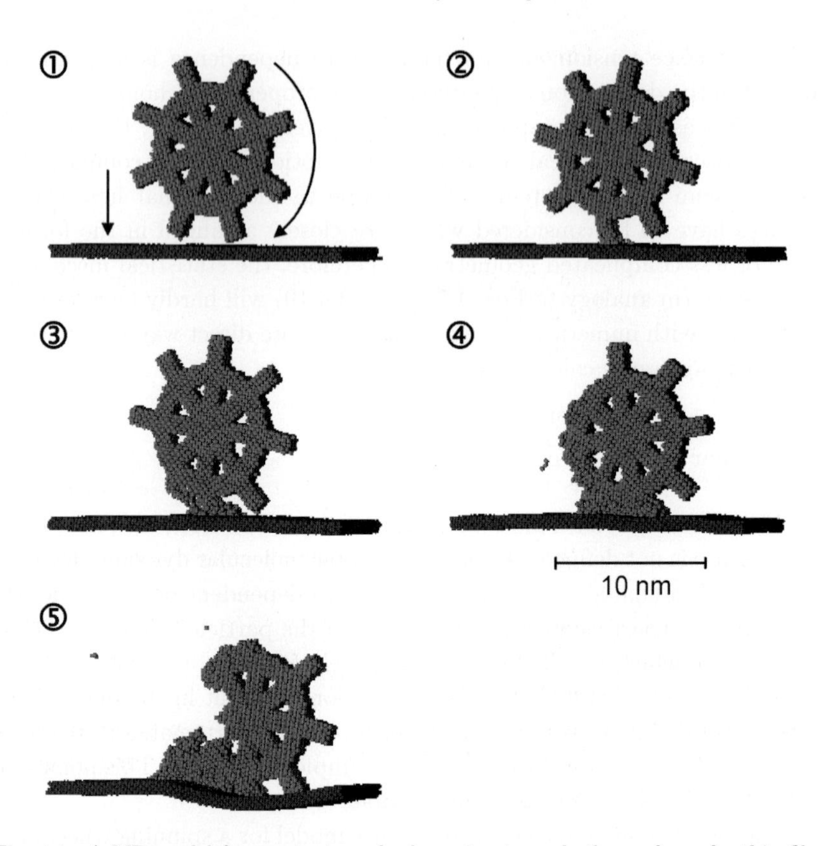

Fig. 4.4 A MD model for a spinning wheel moving towards the surface of a thin film. Both the film and the wheel consist of aluminium atoms. The wheel rotates at room temperature with about 10^{12} revolutions per second. It can be clearly seen that friction effects emerge in form of complex processes as soon as the wheel contacts the surface. In this case, it leads to the destruction of the wheel.

structure and on the harmonic approximation. Due to the considerable anharmonicities and the strongly disturbed structure of nanosystems — even far below the melting temperature — statistical mechanics and the theory of liquids [Rice, Gray, 1965; Lucas, 1991] are also of particular interest. In this context, MD calculations play a major role and, obviously, up to now there is no alternative to this method.

Though visualization is an important analysis instrument, numerical methods are still needed in many cases for quantitative characterizations of the huge amount of MD data. Here, the following sections provide a compilation — including brief derivations — of the most important quantities.

4.4.1 *Ensemble Theory*

The dynamic state of a classical mechanical many-particle system is completely defined, if the generalized co-ordinates $q_i(t)$ and momenta $p_i(t)$ are known at a specific time t. Therefore, the set (q_i, p_i), $i = 1, \cdots, 3N$ can be considered as *micro-state* of the system. This is a point in a $6N$–dimensional space which is the so-called classical *phase space.* The evolution with time of the system follows a curve $(q_i(t), p_i(t))$ in the *phase space* (the *phase space trajectory*) that is determined by the *Hamiltonian equations of motion* (Eq. 3.1). A simple example for a trajectory has already been illustrated in Fig. 3.10.

But there is an alternative point of view: instead of considering *one* system which changes from microstate to microstate, it is possible to consider *many* systems at the same time, all corresponding to the same *Hamiltonian* but each to a *different* microstate. Such collections of systems are called *statistical ensembles* [Schommers, 1986; Friedman, 1985; Greiner *et. al*, 1993].

In general, the probability for finding a macroscopic system in a certain microstate is not equally distributed. Therefore, the states have to be weighted by a probability density function $\rho(q_i, p_i)$ — the so-called *phase space density.* Since this density strongly depends on the environment or boundary conditions, respectively, there are different expressions according to the specific situations which are classified in Table 4.1.

Table 4.1 Statistical ensembles and phase space densities [Schommers, 1986; Greiner *et. al*, 1993].

Ensemble	Density	Conditions
Micro-canonical	$\rho(q_i, p_i) = \begin{cases} \rho_0 & E \leq H(q_i, p_i) \leq E + \triangle E \\ 0 & \text{otherwise} \end{cases}$	isolated system $N, V, E = \text{const.}$
Canonical	$\rho(q_i, p_i) = \dfrac{\exp\left(-\frac{H(q_i,p_i)}{k_B T}\right)}{\int \exp\left(-\frac{H(q_i,p_i)}{k_B T}\right) d^{3N}q\, d^{3N}p}$	closed, isothermal system $N, V, T = \text{const.}$
Grand Canonical	$\rho(q_i, p_i) = \dfrac{\exp\left(-\frac{H(q_i,p_i)-\mu N}{k_B T}\right)}{\int \exp\left(-\frac{H(q_i,p_i)-\mu N}{k_B T}\right) d^{3N}q\, d^{3N}p}$	open, isothermal system $V, T, \mu = \text{const.}$ (μ is the chemical potential)

Now, the basis of the *ensemble theory* is the assumption that any measurable thermodynamic quantity of state f of a macroscopic system can be described by mean values of the according *statistical ensemble quantity* $f(q_i, p_i)$:

$$\langle f \rangle_e = \frac{\int f(q_i, p_i)\rho(q_i, p_i)\, d^{3N}q\, d^{3N}p}{\int \rho(q_i, p_i)\, d^{3N}q\, d^{3N}p} . \tag{4.11}$$

As can be clearly seen from Eq. 4.11 in connection with Table 4.1, in most cases the expression for the mean values of interest becomes rather complicated unless far-reaching approximations are applied. MD calculations, however, enable the direct determination of measurable quantities without approximations in a simple way. This is due to the fact that the MD data automatically represents a statistical ensemble according to the model design. As already described in the previous chapter, most often this is a *micro-canonical* ensemble, but *canonical* or *grand canonical* ensembles may be generated as well. Thus, with the help of MD calculations, Eq. 4.11 takes the simple form:

$$\langle f \rangle = \langle f \rangle_e = \frac{1}{N} \sum_{i=1}^{N} f(\mathbf{r}_i, \mathbf{v}_i) . \tag{4.12}$$

In Eqs. 3.21 and 3.22, for example, this has already been used without discussion.

To average an ensemble, the time dependency of the trajectory plays no role. Instead, each point in phase space is weighted with the density function according to the probability of the occurrence of the specific microstate. Since macroscopic quantities are independent of time in thermodynamic equilibrium, they could be expressed in principle as time averages of single particle quantities along the trajectory by

$$\langle f \rangle_t = \bar{f} = \lim_{\mathcal{T}\to\infty} \frac{1}{\mathcal{T} - \tau_0} \int_{\tau_0}^{\mathcal{T}} f\left(q_i(t), p_i(t)\right)\, dt . \tag{4.13}$$

In theory, the time average $\bar{f}$ would be identical, for example, to the micro-canonical ensemble average $\langle f \rangle_e$, if the trajectory passes each point of the energy surface $H(q_i, p_i) = E$ at least once, but in equal numbers of time. This is the so-called *ergodic hypothesis* [Farquhar, 1964]. Though never exactly fulfilled within complex systems, it may be used for an alternative computation of mean values (*quasi ergodic hypothesis*).

Again, in contrast to theoretical considerations the application of time averages in the case of MD calculations is easy, because time-dependent solutions of the *Hamiltonian* are available for each particle. It has been shown [Schommers, 1986] that the *quasi ergodic hypothesis* is fulfilled in a good approximation already for a few thousand calculation steps, and therefore, within MD calculations the infinite limit in Eq. 4.13 can be replaced by $\tau \rightarrow \tau_e$, where τ_e varies between 10^{-12} and 10^{-11} seconds depending on the MD model.

4.4.2 *Pair Correlation Function*

Within statistical mechanics molecular distribution functions describe the probability of the occurrence of a particular arrangement of atoms in equilibrium. In a *statistical ensemble* of N particles the *pair distribution* $g_2(\mathbf{r}_1, \mathbf{r}_2)\, d\mathbf{r}_1 d\mathbf{r}_2$ is proportional to the probability of finding a particle at $\mathbf{r}_2$ in a volume element $d\mathbf{r}_2$, if, at the same time, there is a particle at $\mathbf{r}_1$ in the volume element $d\mathbf{r}_1$ (independent of where the remaining $N-2$ molecules are located).

For a homogeneous liquid in thermal equilibrium $g_2(\mathbf{r}_1, \mathbf{r}_2)\, d\mathbf{r}_1 d\mathbf{r}_2$ cannot depend on the choice of $\mathbf{r}_1$ and therefore, it depends on the difference $\mathbf{r} = \mathbf{r}_1 - \mathbf{r}_2$ alone. Further, since the liquid is macroscopically an isotropic body, the direction of $\mathbf{r}$ is unimportant. Thus, g_2 depends only on the magnitude of the distance $r = |\mathbf{r}|$. In this way the *pair correlation function* $g(r) = g_2(r)$ is defined as the probability of finding a particle at the distance r, if there is a particle at the origin. Usually the pair correlation function is normalized to unity at large r.

In a mono-atomic system with particle interactions described by a pair potential $v(r)$, the statistical mechanics expression for $g(r)$ is given by [Schommers, 1986]:

$$g(r) = V^2 \frac{\int d\mathbf{r}_3 \cdots \int d\mathbf{r}_N \, \exp\left[-\frac{1}{k_B T} \frac{1}{2} \sum_{\substack{i,j=1 \\ i \neq j}}^{N} v\left(|\mathbf{r}_i - \mathbf{r}_j|\right)\right]}{\int d\mathbf{r}_1 \cdots \int d\mathbf{r}_N \, \exp\left[-\frac{1}{k_B T} \frac{1}{2} \sum_{\substack{i,j=1 \\ i \neq j}}^{N} v\left(|\mathbf{r}_i - \mathbf{r}_j|\right)\right]} \,. \tag{4.14}$$

As can be seen clearly, the determination of the pair correlation function $g(r)$ from the pair potential $v(r)$ by Eq. 4.14 is only possible, if simplifying assumptions are applied on the solution of the N and $(N-2)$-fold integrals. The approximations, produced by such assumptions, are more or less uncontrolled and lead to no reliable results. On the other hand, the determination of $g(r)$ — based on results of MD calculations — can be done without any uncertainty. The above mentioned probability meaning of $g(r)$ is directly converted into an algorithm which is formally given by

$$g(r) = \frac{1}{\varrho} \frac{n(r, \triangle r)}{4\pi r^2 \triangle r} , \tag{4.15}$$

where ϱ is the macroscopic density (N/V) and $n(r, \triangle r)$ is the number of particles within the spherical shell that is given by the radii r and $r + \triangle r$ around an arbitrary particle. The pair correlation function may be applied to surface structures, too.

Then Eq. 4.15 changes to

$$g_{2d}(r) = \frac{1}{\sigma} \frac{n_{2d}(r, \triangle r)}{2\pi r \triangle r} , \tag{4.16}$$

where σ is the two-dimensional macroscopic density in a layer and $n_{2d}(r, \triangle r)$ denotes the number of particles in the annulus (radii r and $r + \triangle r$) around an arbitrary particle within the considered plane.

Typical forms of the pair correlation function are illustrated in Figs. 4.5 and 4.6 for the example of bulk melting and the pre-melting effect on surfaces [von Blanckenhagen, Schommers, 1987; Schommers, 1986; Schommers *et. al*, 1995].

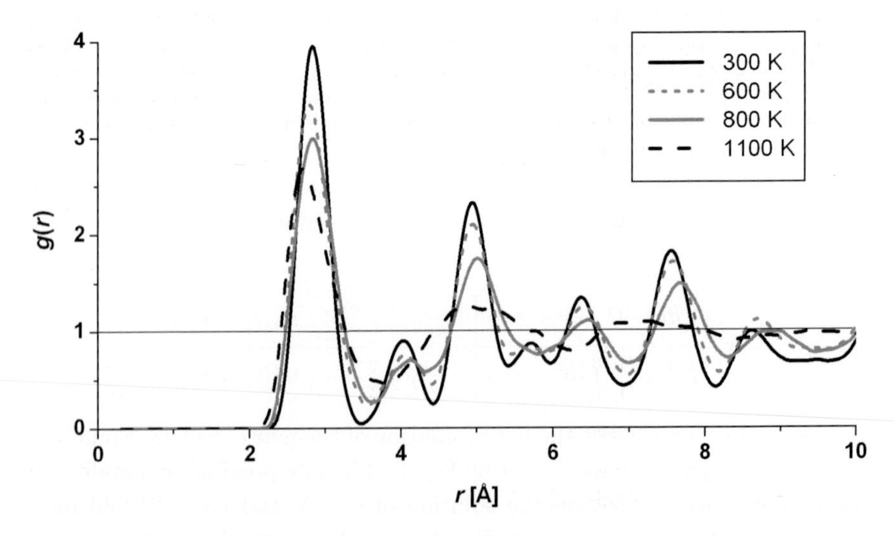

Fig. 4.5 Bulk study with 864 aluminium atoms: The pair correlation function is plotted for different temperatures. The transition from solid to liquid state can be clearly seen (the melting temperature of aluminium is 933 K).

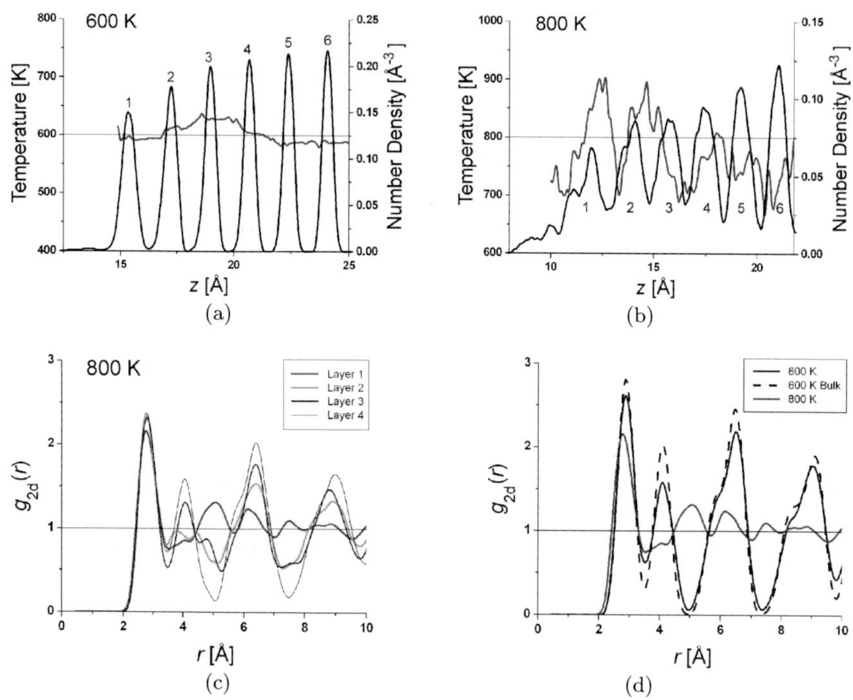

Fig. 4.6 Pre-melting effect on (001) aluminium surfaces: In the present case, the model consists of 2400 atoms arranged in 12 layers. Figures (a) and (b) show the temperature and density profiles perpendicular to the surface (along the z-axis) for mean temperatures of 600 K and 800 K, respectively. The layers are numbered from 1 to 6 starting with the outermost. As can be seen, the temperature fluctuations are stronger in the case of the 800 K study [Fig. (b)], but remain below the melting point of aluminium. The two-dimensional pair correlation functions are plotted in Fig. (c) for the first four layers (800 K study). Obviously, the surface layer shows liquid behavior well below the melting point (pre-melting effect). To verify the result, the pair correlation functions of the surface layers of the 600 K and 800 K studies are compared with a bulk layer [Fig. (d)].

4.4.3 *Mean-Square Displacement*

Characterization of diffusion processes necessitates the determination of the diffusion coefficient D. One possibility to derive D is the computation of the *mean-square displacement* $\langle \mathbf{r}_2(t) \rangle$ as a function of time. The *mean-square displacement* is a measure for the particle mobility and it is calculated by [Schommers, 1986; Egelstaff, 1967]

$$\langle \mathbf{r}^2(t) \rangle = \frac{1}{N} \sum_{i=1}^{N} [\mathbf{r}_i(t) - \mathbf{r}_i(0)] \ . \tag{4.17}$$

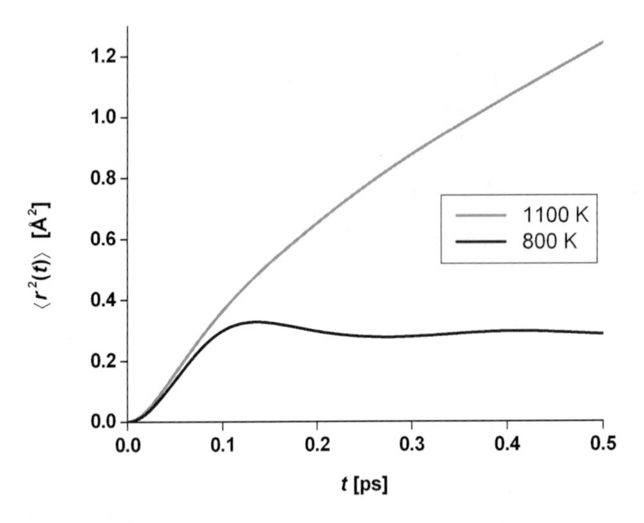

Fig. 4.7 Bulk study with 864 aluminium atoms: The mean-square displacement is plotted for 800 K (solid) and 1100 K (liquid). (The melting temperature of aluminium is 933 K).

With the relation

$$\lim_{t\to\infty} \left\langle \mathbf{r}^2(t) \right\rangle = 2n_d\, t + \text{const.} \tag{4.18}$$

the diffusion constant D can be determined with help of MD calculations over a sufficiently long time period.

For the calculation with Eq. 4.17 particles may be selected from specific regions like planes on or near the surface, and further, by the use of only specific components of the particle co-ordinates $\mathbf{r} = [x, y, z]$ the *mean-square displacement* as well as D may be calculated parallel or perpendicular to a considered direction. In such cases the constant n_d in front of t in Eq. 4.18 has to be adapted according to the dimensionality, e.g. in two dimensions the constant is 2 in three dimensions, it is 3 [Boisvert, Lewis, 1997].

Figure 4.7 shows the results of a liquid (1100 K) and solid (800 K) aluminium bulk study. While the *mean-square displacement* of the solid takes a horizontal course, the slope of the liquid leads to a diffusion coefficient — according to Eq. 4.18 — of about $2.9{\cdot}10^{-5}$ cm^2/s. The experimental value is $3 \cdot 10^{-5}$ cm^2/s [Ludwig, 1969].

In [Schommers *et. al*, 1995], the pre-melting effect on (110) aluminium surfaces has been discussed. Here, the diffusion constant parallel to the outermost layer at 810 K has been determined to be $9.6 \cdot 10^{-5}$ cm^2/s, which

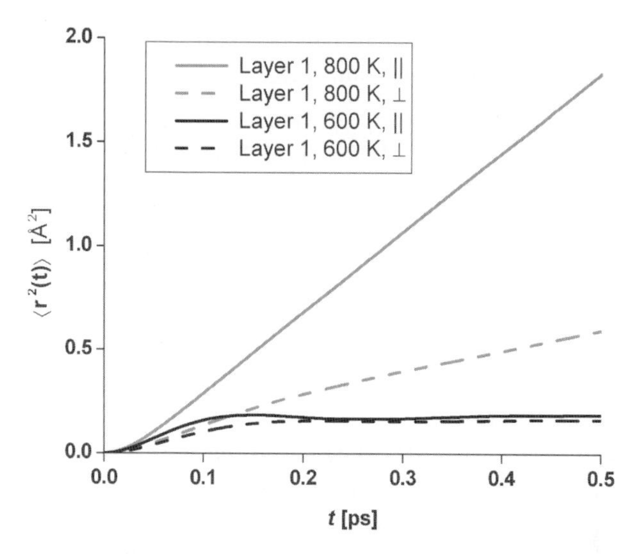

Fig. 4.8 Pre-melting effect on (001) aluminium surfaces. In the present case, the model consists of 2400 atoms arranged in 12 layers. The mean-square displacements parallel and perpendicular to the outermost layer are plotted for 600 K and 800 K.

is distinctly larger than that of the liquid aluminium. Figure 4.8 illustrates the results for (001) surfaces. At 600 K, the *mean-square displacements* parallel as well as perpendicular to the outermost layer (layer 1) take the form of a solid. At 800 K, however, one can see quite clearly that the outermost layer shows the characteristics of a liquid. The diffusion constant parallel to the surface is $9.65 \cdot 10^{-5}$ cm^2/s and the perpendicular one is $2.5 \cdot 10^{-5}$ cm^2/s.

4.4.4 *Velocity Auto-Correlation Function*

Another possibility to derive the diffusion constant D comes by the application of the *velocity auto-correlation function* $\psi(t)$. The *velocity auto-correlation function* is defined by [Schommers, 1986; Egelstaff, 1967]

$$\psi(t) = \frac{\langle \mathbf{v}(0)\mathbf{v}(t)\rangle}{\langle \mathbf{v}(0)^2\rangle} . \tag{4.19}$$

With the velocities $\mathbf{v}_i(t), i = 1, \cdots, N$, calculated by MD, $\psi(t)$ can be computed simply by

$$\psi(t) = \frac{\sum_{i=1}^{N} \mathbf{v}_i(0)\mathbf{v}_i(t)}{\sum_{i=1}^{N} \mathbf{v}_i(0)^2} . \tag{4.20}$$

The *mean-square displacement* can be expressed by the *velocity auto-correlation function* in the following way:

$$\langle \mathbf{r}^2(t) \rangle = \frac{2n_d k_B T}{m} \int_0^t (t-\tau)\psi(\tau)\, d\tau \,. \tag{4.21}$$

Then, with Eq. 4.18 the diffusion constant D may be derived alternatively by

$$D = \frac{k_B T}{m} \int_0^\infty \psi(t)\, dt \,, \tag{4.22}$$

but it should be noted that the numerical determination of D according to Eq. 4.18 is more precise.

Here again, with the MD data, it is easy to determine direction-dependent diffusion constants simply by selecting the particles within a desired area, e.g. in a certain layer, and by using the specific vector components of the particle velocities $\mathbf{v} = [v_x, v_y, v_z]$ or by transformation of the co-ordinate system [Schommers, 1986; Schommers, 1987].

Figure 4.9 shows the *velocity auto-correlation functions* for solid (800 K) and liquid (1100 K) aluminium. The minimum of $\psi(t)$ in the case of the solid is clearly more pronounced compared to the liquid.

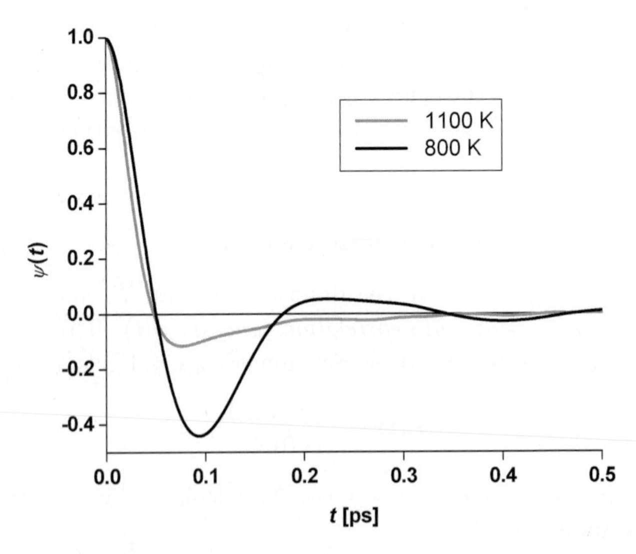

Fig. 4.9 Bulk study with 864 aluminium atoms: The velocity auto-correlation function is plotted for 800 K (solid) and 1100 K (liquid). (The melting temperature of aluminium is 933 K).

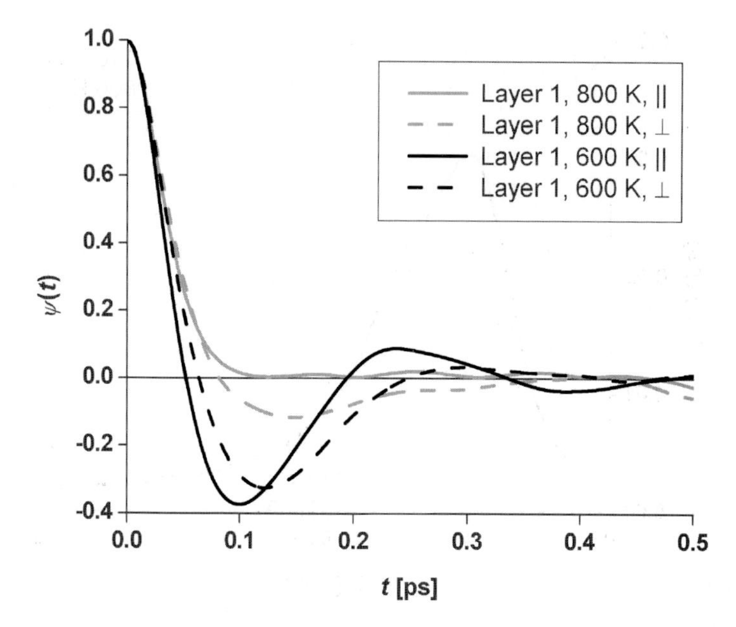

Fig. 4.10 Pre-melting effect on (001) aluminium surfaces: In the present case, the model consists of 2400 atoms arranged in 12 layers. The velocity auto-correlation functions parallel and perpendicular to the outermost layer are plotted for 600 K and 800 K.

The pre-melting effect is illustrated in Fig. 4.10 for the outermost layer (layer 1) at 600 K and 800 K of a (001) aluminium surface.

4.4.5 *Generalized Phonon Density of States*

The *Fourier transform* of $\psi(t)$ leads to a frequency spectrum $f(\omega)$ which, in the case of harmonic solids, represents the frequency spectrum $g(\omega)$ of the normal modes, i.e. the *phonons*. Thus, for anharmonic systems, like nano-clusters or surfaces, the *Fourier transform* $f(\omega)$ of the *velocity auto-correlation function* can be considered as the *generalized phonon density of states* [Schommers, 1986]

$$f(\omega) = \frac{1}{N_n} \frac{2}{\pi} \int_0^\infty \psi(t) \cos \omega t \, dt \,, \tag{4.23}$$

where N_n is used for the normalization to unity and is given by

$$N_n = \frac{2}{\pi} \int_0^\infty \int_0^\infty \psi(t) \cos \omega t \, dt \, d\omega = 1 \,. \tag{4.24}$$

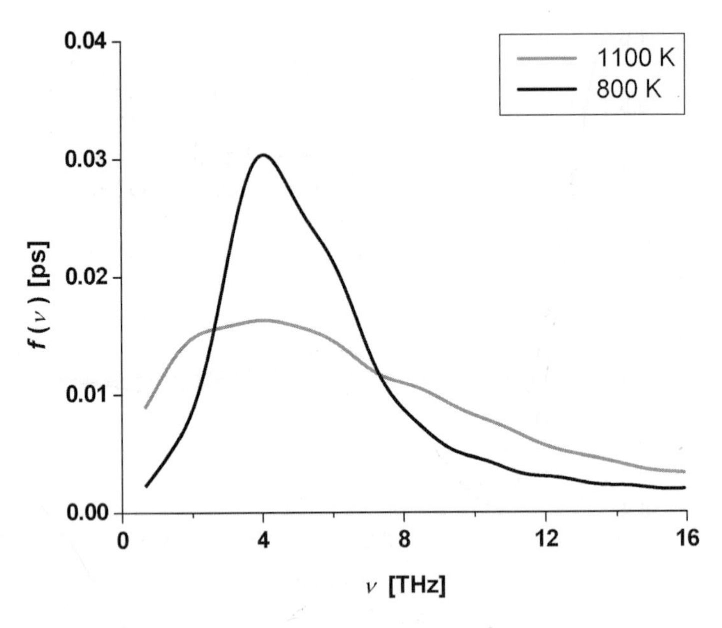

Fig. 4.11 Bulk study with 864 aluminium atoms: The generalized phonon density of state is plotted for 800 K (solid) and 1100 K (liquid). (The melting temperature of aluminium is 933 K).

The comparison of the Eqs. 4.22 and 4.23 finally leads to the following formulation for the diffusion constant D:

$$D = N_n \frac{\pi}{2} \frac{k_B T}{m} f(\omega = 0) , \tag{4.25}$$

which is equivalent to Eq. 4.22.

Figure 4.11 shows the *generalized phonon density of states* for solid (800 K) and liquid (1100 K) aluminium. It can be seen that the maximum of the frequency spectrum is more pronounced in the case of the solid.

The results for the pre-melting effect of the outermost layer (layer 1) are illustrated in Fig. 4.12. The *phonon densities of state* of the solid (001) aluminium surface (600 K) show well pronounced maxima, whereas the spectrum of the phonon density perpendicular to the surface is shifted toward lower frequencies by about 0.7 THz.

In the case of pre-melting (at 800 K) only the spectrum perpendicular to the surface shows a recognizable maximum at an even lower frequency.

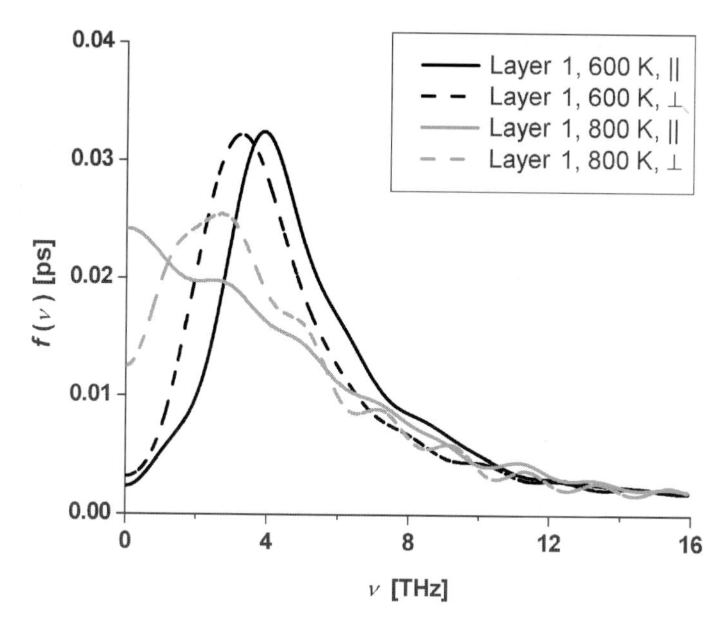

Fig. 4.12 Pre-melting effect on (001) aluminium surfaces: In the present case, the model consists of 2400 atoms arranged in 12 layers. The generalized phonon densities of state parallel and perpendicular to the outermost layer are plotted for 600 K and 800 K.

4.4.6 *Structure Factor*

The *structure factor* $S(\mathbf{q})$ is defined by [Egelstaff, 1967]

$$S(\mathbf{q}) = 1 + \rho \int_V e^{i\mathbf{qr}} \left[g(r) - 1\right] d\mathbf{r} \ , \tag{4.26}$$

and for $\mathbf{q} \neq 0$ $S(\mathbf{q})$ is proportional to the neutron cross-section. In isotropic materials, e.g. liquids, it is a function of the magnitude of $\mathbf{q}$ only, since rotation of the target does not alter the intensity of scattered radiation. Thus, $S(q)$ can be derived in principle by *Fourier transform* of the *pair correlation function* $g(r)$. But due to cut-off effects in the calculation of $S(q)$ by Eq. 4.26, the direct numerical *Fourier* transform is not appropriate for the determination of $S(q)$. One possibility to improve the application of Eq. 4.26 is to apply a smoothing algorithm on $g(r)$. Another method is the approach from the *intermediate scattering function* [Salacuse *et. al*, 1986].

The *intermediate scattering function* $I(\mathbf{q}, t)$ is defined as the self-correlation of the *Fourier* transform $\rho(\mathbf{q}, t)$ of the microscopic number density $\rho(\mathbf{r}, t)$. Considering a system of N particles with the positions $\mathbf{r}_i(t)$,

$i = 1, \cdots, N$, the relations are

$$\rho(\mathbf{r}, t) = \sum_{i=1}^{N} \delta\left[\mathbf{r} - \mathbf{r}_i(t)\right] , \tag{4.27}$$

$$\rho(\mathbf{q}, t) = \sum_{i=1}^{N} \exp\left[i\mathbf{q}\mathbf{r}_i(t)\right] , \tag{4.28}$$

$$I(\mathbf{r}, t) = \frac{1}{N}\langle \rho(-\mathbf{q}, 0)\rho(\mathbf{q}, t)\rangle \tag{4.29}$$

$$= \frac{1}{N}\left\langle \sum_{k=1}^{N}\sum_{\substack{i=1\\ i\neq k}}^{N} \exp\left[i\mathbf{q}\left\{\mathbf{r}_i(t) - \mathbf{r}_k(0)\right\}\right]\right\rangle ,$$

where $\langle \cdots \rangle$ represents the equilibrium ensemble average. $I(\mathbf{q}, t)$ is related to the *dynamic structure factor* $S(\mathbf{q}, \omega)$ by *Fourier transformation*:

$$S(\mathbf{q}, \omega) = \mathfrak{F}\{I(\mathbf{q}, t)\} , \tag{4.30}$$

which can be measured directly by neutron scattering experiments.

Then, $S(\mathbf{q}, \omega)$ is related to $S(q)$ by [Egelstaff, 1967]

$$S(\mathbf{q}) = \int_0^{\infty} S(\mathbf{q}, \omega)\, d\omega . \tag{4.31}$$

While MD calculations take place within a simulation space of box lengths L_x, L_y, L_z the non-zero $\mathbf{q}$ values are restricted to

$$q = \left[\frac{l_x}{L_x}, \frac{l_y}{L_y}, \frac{l_z}{L_z}\right] , \tag{4.32}$$

where $l_x, l_y, l_z = 0, \pm 1, \pm 2, \cdots$.

Now, with the restriction of Eq. 4.32, a set of $\mathbf{q}$ values, which magnitudes $q = |\mathbf{q}|$ are in the range $[q, q + \mathrm{d}q]$, can be determined numerically.

Finally $S(q) = I(q, t = 0) = \langle I(\mathbf{q}, t = 0)\rangle_q$ can be obtained by use of Eq. 4.29 as an average over all $\mathbf{q}$ values from the according sets.

The analytically performed averaging over the sets of $\mathbf{q}$ values in Eq. 4.29 leads to the following expression for the *structure factor*, which can be used alternatively [Salacuse *et. al*, 1986]:

$$S(q) = \frac{1}{N}\sum_{k=1}^{N}\sum_{\substack{i=1\\ i\neq k}}^{N} \frac{\sin\left[iq\left|\mathbf{r}_i - \mathbf{r}_k\right|\right]}{q\left|\mathbf{r}_i - \mathbf{r}_k\right|} . \tag{4.33}$$

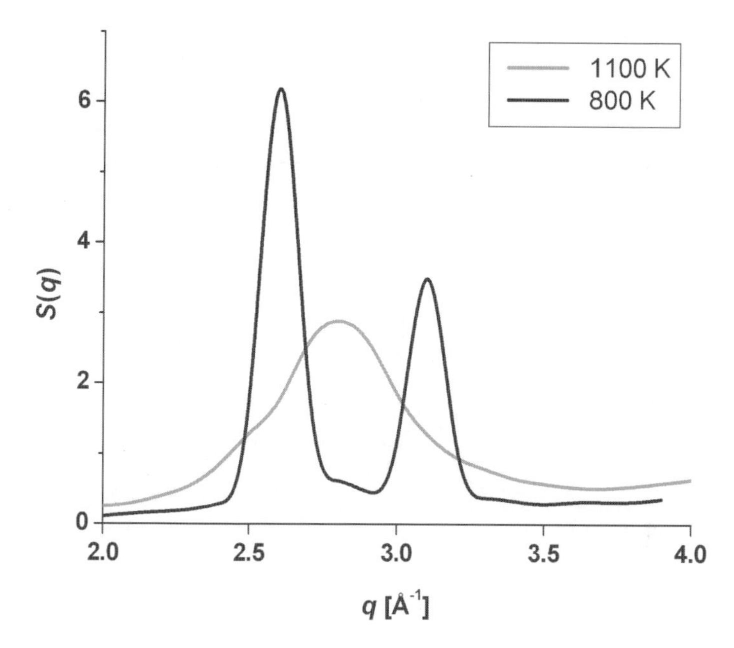

Fig. 4.13 Bulk study with 864 aluminium atoms: The structure factor is plotted for 800 K (solid) and 1100 K (liquid). (The melting temperature of aluminium is 933 K).

Figure 4.13 shows the structure factors derived according to the Eqs. 4.31 and 4.32 for solid (800 K) and liquid (1100 K) aluminium. The *structure factor* of the solid shows two peaks, one at 2.6 Å^{-1} and a smaller one at 3.1 Å^{-1}. Such double peaks are typical for fcc structured crystals.

In the liquid state, the *structure factor* has the form of a rather broad peak with a maximum at 2.8 Å^{-1}.

Figure 4.14 illustrates the situation for a (001) aluminium surface. The two-dimensional structure factors are plotted for the first three layers at 800 K and for the outermost layer at 600 K.

As can be seen, crystalline structures are indicated by double peaks of the *structure factor.* Compared to the three-dimensional case, here the first peak appears already at 2.2 Å^{-1}.

At 600 K, the outermost layer is still in crystalline state and at 800 K the structure factors of the third and second layer show double peaks which indicates crystalline structure too. But the outermost layer is in a liquid state at 800 K. This can be clearly seen from the flat form of the structure factor where only the former first peak is weakly recognizable.

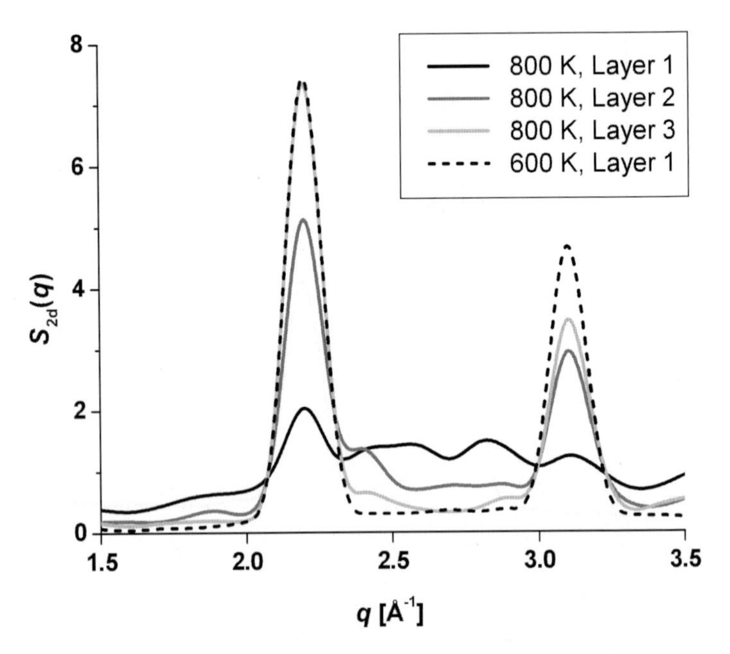

Fig. 4.14 Pre-melting effect on (001) aluminium surfaces: In the present case, the model consists of 2400 atoms arranged in 12 layers. The two-dimensional structure factors are plotted for the first three layers (800 K study). Obviously, the surface layer shows liquid behavior well below the melting point (pre-melting effect). To verify the result, the structure factor of the surface layers of the 600 K study is plotted in addition.

4.4.7 *Additional Remarks*

To reduce the fluctuations occurring with the calculation of the above mentioned time-dependent correlation functions, it is necessary to take the time average for several calculation steps. In most cases, the average over about 100 time steps is a sufficiently exact representation of the ensemble average.

Then, with n_t being the number of time averages, e.g. Eqs. 4.17 and 4.20 — for the *mean-square displacement* and the *velocity auto-correlation function* take the form

$$\langle \mathbf{r}^2(t) \rangle = \frac{1}{n_t}\frac{1}{N}\sum_{j=1}^{n_t}\left[\mathbf{r}_i(t+t_j) - \mathbf{r}_i(t_j)\right]^2 \tag{4.34}$$

and

$$\psi(t) = \frac{1}{n_t}\frac{1}{N}\sum_{j=1}^{n_t}\frac{\sum_{i=1}^{N}\mathbf{v}_i(t_j)\mathbf{v}_i(t_j+t)}{\sum_{i=1}^{N}\mathbf{v}_i^2(t_j)}\,. \tag{4.35}$$

Chapter 5

Nano-Engineering — Studies and Conclusions

In general, nano-machines can be defined as systems in the nanometer realm with moving parts. [Drexler, 1992], and others too (e.g. [Han *et. al*, 1997]), have already shown numerous designs of nano-machines that are exclusively based on carbon structures. Due to their covalent bonds, the benefit of such organic compounds is a rather high stability — especially with diamondoid structures or arrangements similar to fullerenes (typical forms are C_{60}-balls or nano-tubes). To determine stable configurations of such structures an extensive use of *molecular mechanics* or even *ab initio* methods is necessary.

In the frame of the present monograph, however, we would like to consider nano-machines that consist at least partly of metallic compounds. There are several reasons to do this.

First of all, the functionality of nano-machines can be extended significantly due to the electronic properties of metals.

Further, metallic compounds allow complex shaped parts for which models can be set up in a relatively simple way by using computer aided nano-design software as outlined in Sec. 3.1.4 (see, for example, Fig. 3.8).

And finally, in contrast to the anisotropic, covalent bonded organic structures which have to be described by two-, three- and even four-body potentials, we have shown that aluminium (as a representative for a metallic compound) can be modeled with high precision by an isotropic pair potential.

Whether metallic or carbon based structures, the synthesis of real nano-machines has not yet been performed and it is needless to discuss which one is more appropriate to do so. But it is a most interesting topic to perform *molecular dynamics* studies and, therefore, to determine how and under which conditions nano-machines could be working — regardless of the question of how to assemble them experimentally.

5.1 Functional Nanostructures

Before we go into the details of assembling nano-machines, let us first consider design examples of some more or less complex but motionless nanostructures.

As outlined in Sec. 3.1, nano-designing metallic structures may start with a continuous CAD model. Then it is filled up, atom by atom, with the material of choice — in our case with aluminium — according to the perfect lattice structure. The resulting nano-object can be considered as *MD model for* 0 *K*, because the atoms are not in motion. Therefore, we have to perform an *isothermal equilibration process* with a specific temperature according to our design prescriptions. Now, the resulting (isolated) nanostructure is in thermal equilibrium and we should check whether the structure remains stable (it might be melting or dissolve). This can be done by a sufficient number of MD calculation steps, i.e. we have to trace the temporal development of the nanostructure until no further change is detectable. In this way — which can be considered as *computational nano-engineering procedure* — we gain a realistic representation of our initial nano-model.

Figure 5.1 illustrates an arbitrary nano-design example. Here, the MD model at 0 K matches perfectly the shape of the initial CAD model. But, as becomes clear after the equilibration phase, this is not a stable structure at room temperature. During equilibration the atoms have changed their positions at several locations within the structure and as a result the shape of the final stable design is somewhat different from that of the original model. Compared to the strictly cubical shape of the MD model the resulting stable structure now shows bent bars. At the same time, it seems, that the edges are denser and therefore, the side length are slightly shorter.

At this stage, a nano-engineer has to decide whether the nanostructure still meets the design criteria. If this is not the case, he has to redesign the model.

Further nano-designs are shown in Figs. 5.2 and 5.3. In principle, such strange or probably even more complex shaped structures, might play a decisive role in the development of future medical care strategies. To

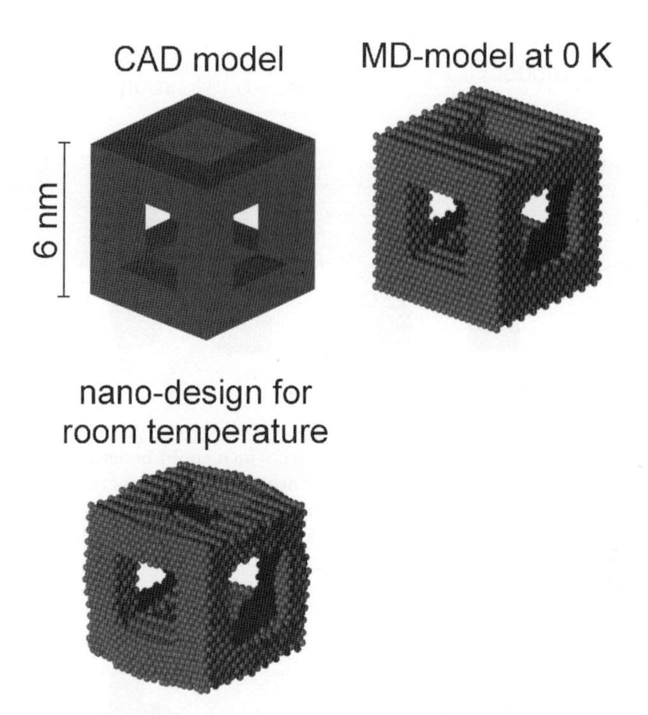

Fig. 5.1 The three steps of nano-design. A continuous CAD model is used to construct the according MD model. After a suitable isothermal equilibration process (in this case at 300 K) the final stable configuration is at disposal. It can be seen quite clearly, that at room temperature the outer shape of the stable structure differs from the initial model. Here, the nanostructure consists of 6820 aluminium atoms.

understand the basic idea, we have to make a short excursus through *molecular biology.*

Generally, human cells may be considered as egg-shaped objects with a medium size of roughly 100 μm (of course, cell shape as well as cell size depend strongly on the cell type, but for the following this simple picture is sufficient for our considerations). The cell surface is covered by "docking stations" (receptors) for a lot of quite different purposes, for example, to control the cell division or for the inter-cell communication, to name only two. Each type of receptor has a specific complex shape which allows only certain specific complex shaped counterparts to dock with. The *growth factor molecule*, for example, is such a counterpart. It consists of 191 amino chains (the simplest form of amino acid is H_2N–CH_2–COOH). The length of another rather complex molecule that plays a role in cell–cell adhesion and signal transduction (E-cadherin) is more than 550 amino

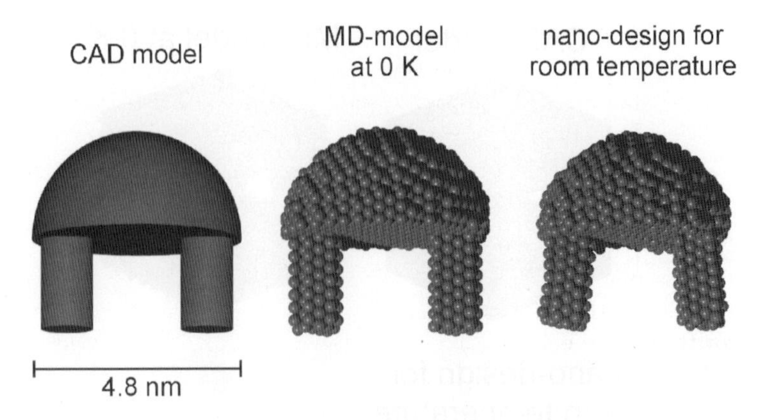

Fig. 5.2 Nano-design for a prototype structure which could be used to perform specific medical tasks. Obviously, at room temperature the stable structure differs somewhat from the original design (both "feet" are bent outward). Depending on the importance of the exact geometrical shape for the work, the structure has to be redesigned. The model consists of 1958 aluminium atoms.

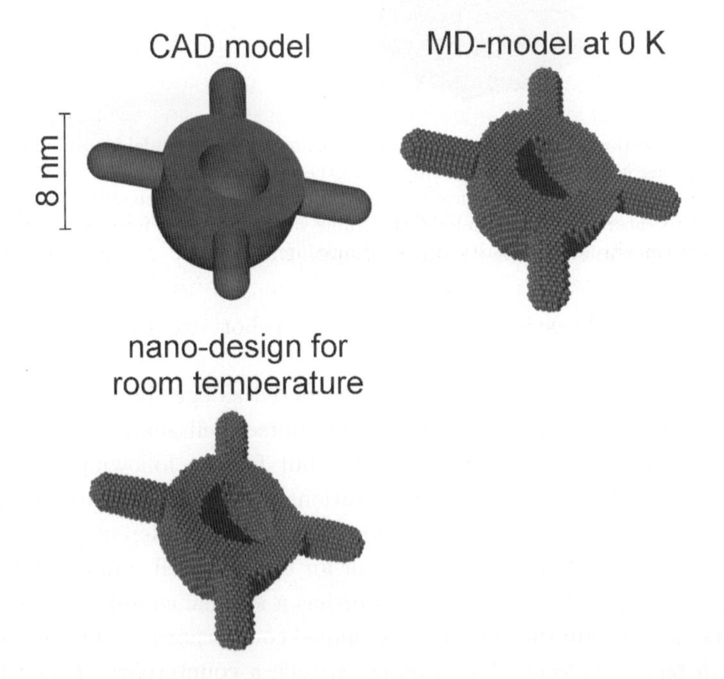

Fig. 5.3 Another model of a prototype nanostructure with possible medical operations. In comparison to the model of Fig. 5.2, the design is larger in size and consists of more rounded parts. Therefore, the final stable structure — which consists of 14235 aluminium atoms — meets the design criteria nearly perfect.

acids. For a comparison, one winding of the DNA strand has a length of 3.4 nm, a diameter of 1 nm, and contains 10 base pairs plus the according polymerized nucleotids of the double helix structure. Each docking process triggers a complicated chain reaction within the cells through enzymes and other signalling molecules which finally cause actions like cell division or cell destruction.

In this way, a lot of important cellular mechanisms are controlled. Unfortunately, there are diseases like cancer and certain kinds of virus infections that interrupt or alter this mechanisms and therefore, uncontrolled cell growth may be enabled or the immune system can be prevented from detecting malfunctioning cells.

During the last decade, a lot of progress has been made in understanding the special mechanisms of such diseases. And in some rare cases this knowledge has even led to the synthesis of artificial tailor-made molecules that can take decisive action against certain miscontrolled cells.

In this context, the nano-designs of Figs. 5.2 and 5.3 may be understood as inorganic prototypes for a specific interaction with the cell control and communication system. Of course, these are rather basic and simplified models. A more realistic design should at least consider the oxidation of the aluminium surfaces and the embedding of some functional organic compounds at the ends of the structures to find out whether metallic structures are suitable at all to perform such tasks.

However, as can be seen clearly, the stability of a nanostructure depends strongly on the size and its outer shape. In the case of the larger design with more rounded parts (Fig. 5.3), the deviations from the MD model are rather small compared to those of the stable structure of Fig. 5.2.

Thinking of today's chip technology, the following question might arise: where is the lower limit of miniaturization in connection with electronic signal processing? Or, in other words, how stable are metallic wires of nanometer size?

The according design example is illustrated in Fig. 5.4. Again, as a first step a very simple model is set up to show the principle. Here, bar shaped wires are arranged horizontally to address matrix points which are located at a (virtual) surface. The connection to the surface is implemented by vertically arranged wires. However, the nano-design for room temperature shows that metallic aluminium structures with quadratic cross-sections of about 1–2 nm^2 can be stable.

As already mentioned, nano-engineering is an iterative process. In the examples given so far, only the structural stability without any interaction

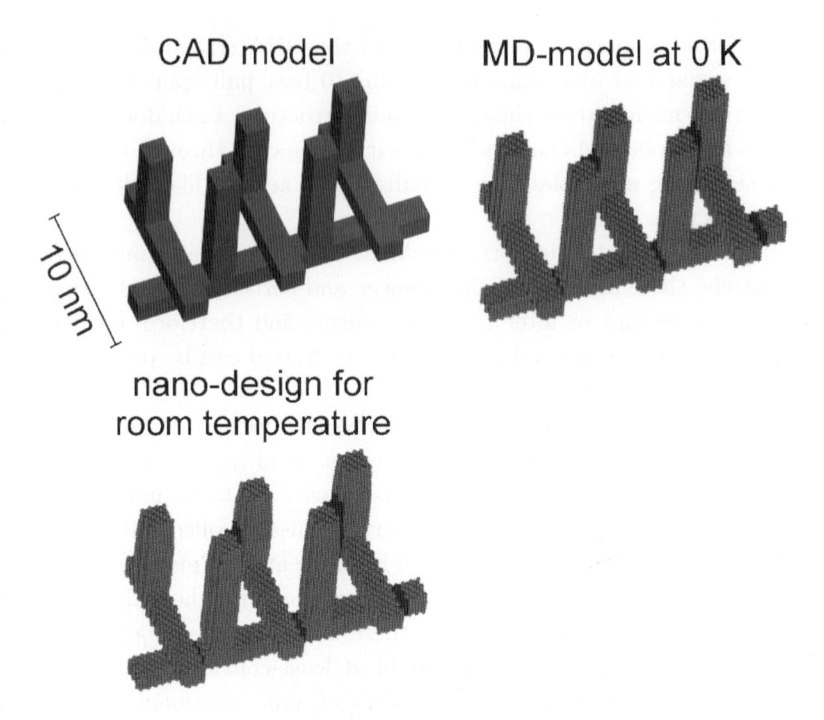

Fig. 5.4 A simplified nano-design of an array of conducting wires. The horizontally arranged wires may be used for (x, y)-addressing a certain point at the surface of the structure (the vertical wires are the connection to the surface). Embedded in an isolated environment such arrays could be used either as a sensor or — in combination with quantum dots — as memory chip or similar electronic device. The model consists of 13304 aluminium atoms.

with the environment has been considered, i.e. the nanostructures has been designed as isolated objects in vacuum. Now, if the (isolated) design meets the prescribed criteria in vacuum, the next engineering step could be, for example, to implement interactions with other structures. In the case of the nano-array in Fig. 5.4, the gaps between the wires could be filled with another material that acts as isolator. In the case of the medical prototypes of Figs. 5.2 and 5.3, the structures could be embedded in an organic liquid to simulate the biological environment of the human organism.

5.2 Nano-Machines

Now we would like to focus on nano-machines. In this case, the use of at least two different materials is unavoidable because we have to deal with movable parts. In our macroscopic world, the design of a simple bearing (which basically can be a drill-hole) and an axle (that can be a cylindrical rod) is not difficult. Both, bearing and axle may consist of the same material. Even if the rod has nearly the same diameter as the drill-hole, the axle is still movable due to the presence of a film of lubricant (water, dirt, air, grease, etc.). If we imagine the same situation at the nanometer realm, both parts would merge together forever.

But there are still other difficulties. Due to the atomic structure, continuous surfaces and shapes are not possible at the nanometer level. Figure 5.5 shows the situation for simple bearings and axles of different diameters.

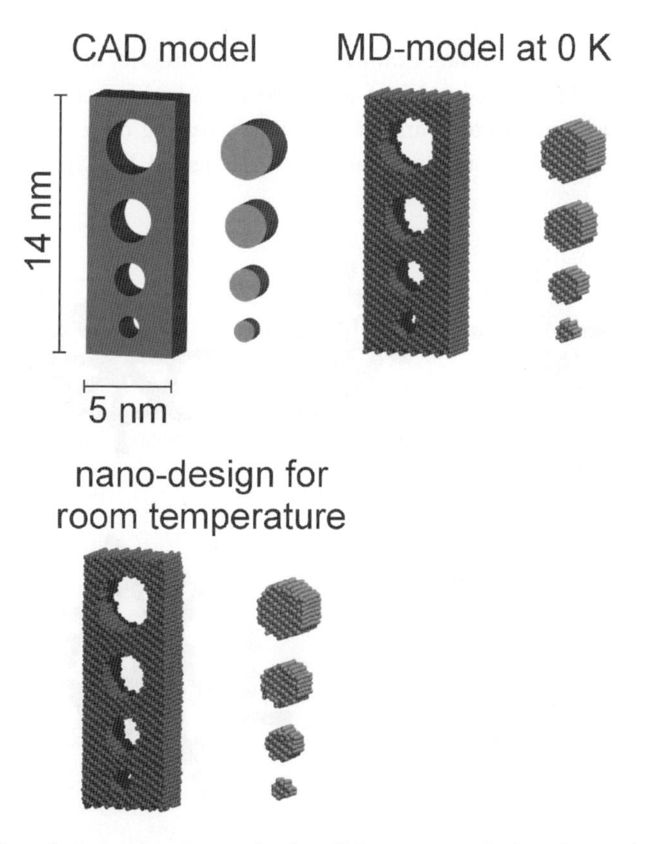

Fig. 5.5 Nano-designing bearings and axles. Tolerances may be kept in a rather narrow range with mechanical engineering, but nano-design allows measures only in certain steps where the step width depends on the according material. Therefore, designing ideal cylindrical shapes and exact fits of axle and bearing is not possible in the nanometer realm.

That is, fits for nano-parts, made of crystalline structured materials, like metals or carbon (diamond), have to be designed rather loose.

Rotating parts are of particular interest in connection with nano-machines. The simulation of a rotating nano-structure is quite simple: the first design step is to create a motionless structure as outlined in the preceding section, then velocity vectors — according to the prescribed revolution speed — are added to each atom before the MD calculation is continued. In this way, centrifugal forces appear automatically as a consequence of the equations of motion — they must not be added to the calculation.

Figure 5.6 shows an example for a nano-wheel of krypton. Here, the structure disintegrates because the appearing centrifugal forces are too strong.

Another, more complex example of such an integrity study is illustrated in Fig. 5.7. This is a model of a simple nano-turbine that consists of a paddle-wheel with an axle and two bearings, resting on a substrate. Here, the propulsion of the turbine could be established by a particle or laser beam.

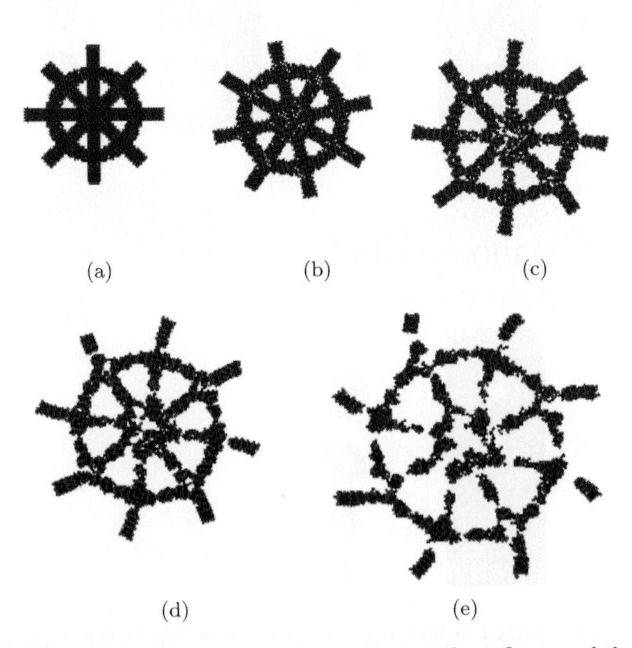

Fig. 5.6 Stability of rotating nanostructures. To test the influence of the revolution velocity ω on the integrity, the structure is set in motion after equilibration. If the resulting centrifugal forces are too large, the structure disintegrates as illustrated in the Parts (a)–(e).

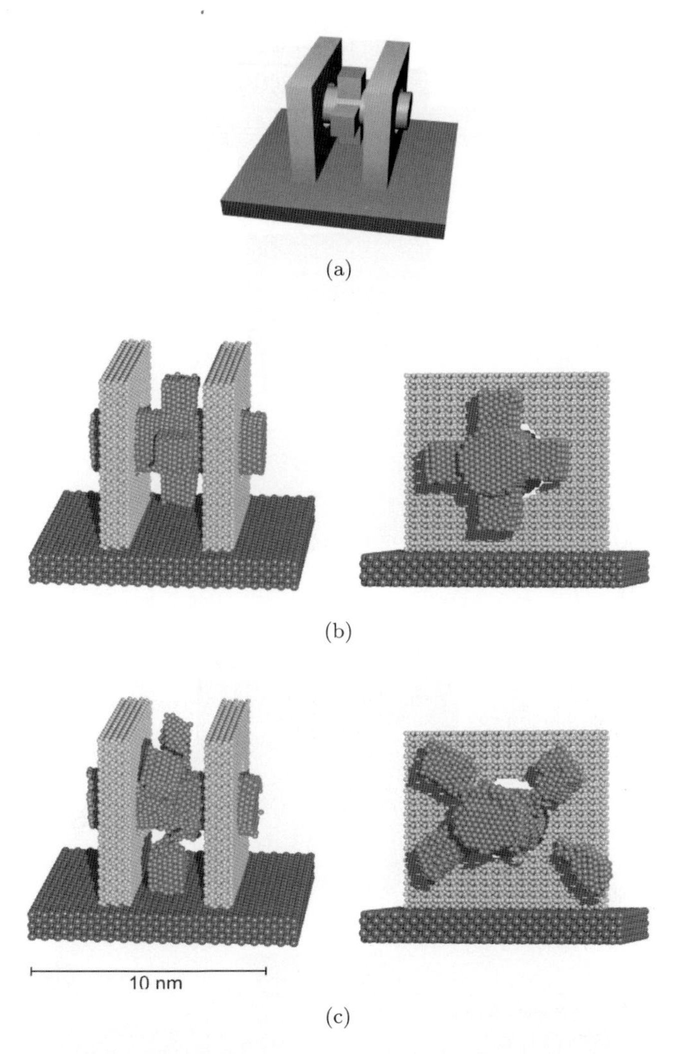

Fig. 5.7 Nano-design of a simple turbine consisting of a paddle-wheel with an axle and two bearings: (a) Mechanical engineering model build with common CAD software. (b) Perspective view and cross-section of the MD model. The paddle-wheel rotates with $5 \cdot 10^{10}$ revolutions per second. (c) The paddle-wheel ruptures, if the number of revolutions is increased to 10^{11} per second. The bearings consist of 4592 silicon atoms, each with ideal shaped surfaces. Due to the strong covalent bonds of silicon the bearings have been treated as rigid objects within the MD calculations. The axle with the paddle-wheel has been assembled with 5226 aluminium atoms and the substrate consists of krypton. Within the MD studies Al–Al interactions have been described by the Schommers potential. For the Al–Si interaction, Eq. 2.19 has been applied. Therefore, the Al–Al potential has been fitted to Buckingham form (Eq. 2.18) and the parameters for silicon have been taken from Table 2.1.

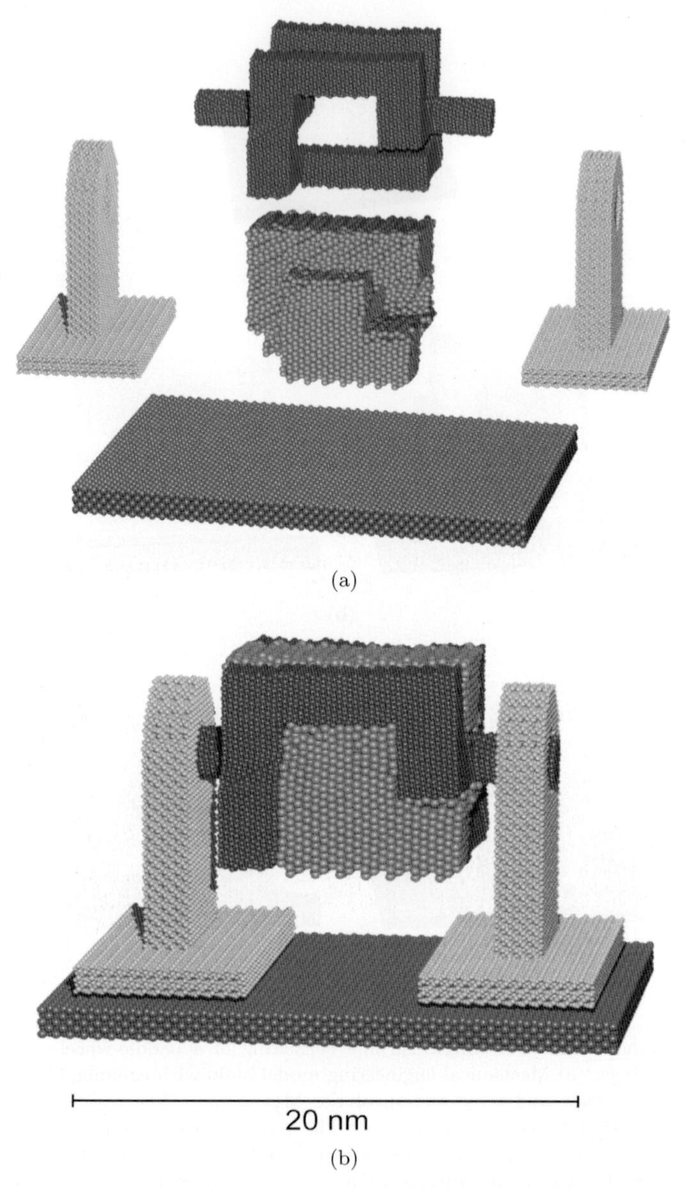

Fig. 5.8 Nano-generator for electric power: (a) Single parts: Winding with axles, isolating rotor kernel, bearings and substrate. (b) The assembled nano-generator rotating with $5 \cdot 10^9$ revolutions per second. The winding consists of 25433 aluminium atoms and the rotor kernel consists of 11341 krypton atoms. The bearings have been treated as rigid objects (silicon) and the interaction potentials for different materials have been derived as described in Fig. 5.7.

When comparing Fig. 5.7(a) with Fig. 5.7(b), the differences between common *mechanical engineering* and *nano-design* become obvious again. While mechanical engineering usually works with sharp edges and smooth surfaces, nano-design is restricted to the use of single atoms which leads to the typical "grainy" appearance of nanostructures. Moreover, tolerances may be kept very small with mechanical engineering, but nano-design allows measures only in certain steps where the step sizes depend on the according material. Therefore, the fit of axle and bearing is rather loose, as can be seen in the cross-sections of Fig. 5.7. However, the MD studies have shown that such nano-turbines could be working stable for revolutions of up to about $5 \cdot 10^{10}$ per second.

The MD model of another principle study is illustrated in Fig. 5.8. It is a simple electric power generator consisting of one winding, stabilized by an isolating kernel which rotates between two bearings. It could be driven

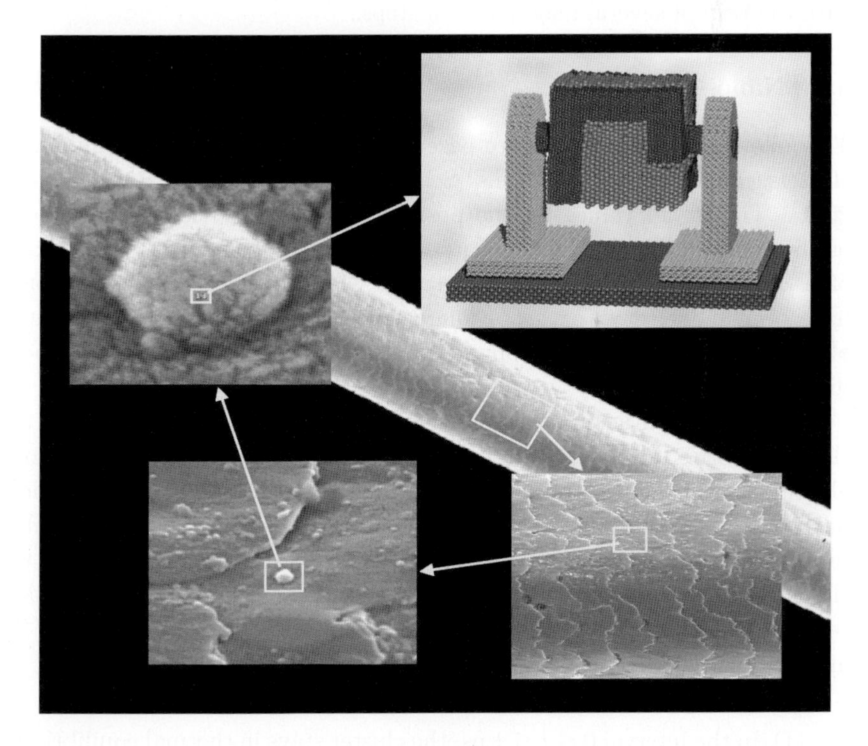

Fig. 5.9 For a better imagination of the proportions, the nano-generator of Fig. 5.8 is illustrated together with a human hair. Since it has a diameter of 80 μm, there could be lined up about 12000 nano-generators around the hair.

either by a nano-turbine similar to that of Fig. 5.7 or by attached paddles to the rotor kernel. MD studies have shown that a stable operation with about $5 \cdot 10^9$ revolutions per second would be possible. Due to the magnetic field of the earth, an electrical voltage of about 10^{-11} V between the axles would be induced.

By a closer look at the surface of the winding and rotor kernel the main problem that occurs by using metals or non-bonded materials for nano-machines becomes obvious. Due to the small size of these parts in addition with temperature effects and/or external forces, there are distortions and dislocations of the perfect crystal lattice structure. This in turn causes slight deformations of the outer shapes. Such structural transformations will be discussed in detail in the following section.

However, in future, nano-machines of any kind could play a major role within many fields of application. Since it is rather difficult to realize the magnitude of nano-scaled objects, Fig. 5.9 tries to illustrate the situation with the help of several magnification steps.

5.3 Nano-Clusters

The following considerations are based on MD calculations for aluminium clusters consisting of 500 atoms [Schommers, Rieth, 1997; Rieth *et. al*, 2001]. The initial values for the positions have been chosen to be those of the bulk structure and the outer shape of the clusters has the form of a cube with (001) oriented surfaces. The inter-atomic interactions have been modeled with the *Schommers* potential using a cut-off radius of 7.35 Å which coincides approximately with the six nearest neighbor distance.

To gain a high accuracy for rather long time periods with respect to the trajectories the 6-*value Nordsieck/Gear predictor-corrector* integration algorithm has been used with a step width of 10^{-15} s. The initial velocities have been set according to a temperature of 50 K followed by an equilibration phase of 2000 time steps (2 ps). In the following, $t = 0$ refers to the end of the equilibration phase, i.e. the time at which the system has reached thermal equilibrium. Figure 5.10 shows the cluster temperature T and the α-function (see Eq. 3.21) as a function of time t.

As can be seen quite clearly, there are three different phases:

(1) In the interval $0 \leq t \leq 1$ ps, the cluster stays in thermal equilibrium ($\alpha = 5/3$) at a mean temperature of 50 K (with the usual, stable fluctuations).

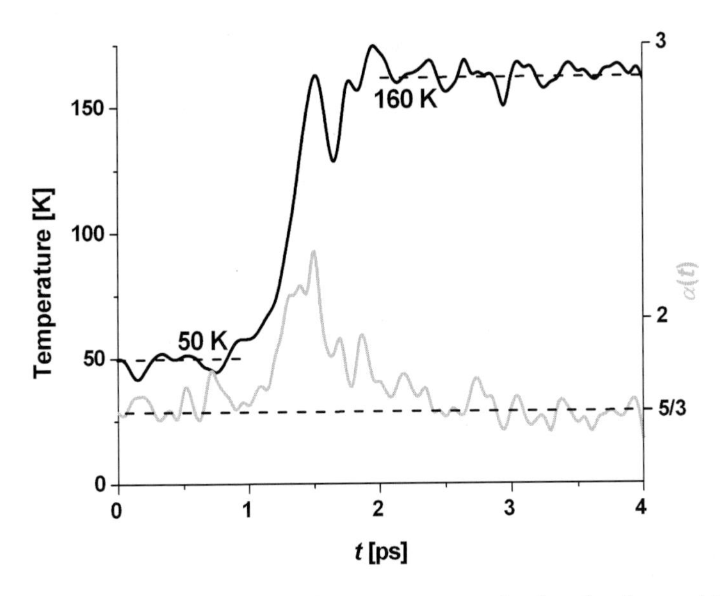

Fig. 5.10 Temporal development of the temperature and α-function for a cubic Al_{500} cluster with (001) surfaces: Initially, the cluster remains in thermal equilibrium at about 50 K for 1 ps ($\alpha = 5/3$ indicates Maxwellian distribution). Then, during a transition phase the cluster temperature increases up to 160 K. During the transition phase of 1 ps the cluster leaves the equilibrium. After this, the mean temperature remains at 160 K for all times and in thermal equilibrium. However, it has to be mentioned that there is no external influence causing the transition.

(2) Within the following interval 1 ps $\leq t \leq$ 2 ps and without external influence a transition takes place during which the temperature rises up to 160 K. Here, the cluster is not in thermal equilibrium ($\alpha > 5/3$).
(3) Finally, for $t > 2$ ps, the cluster temperature remains at a mean value of 160 K, again, in thermal equilibrium.

How can this interesting behavior be interpreted? What is the driving force behind the phenomenon? And, under which conditions does it occur?

5.3.1 *Structural Examinations*

Obviously, the cluster is in a *meta-stable state* within the first phase ($T =$ 50 K) and in a *stable state* within the third phase ($T =$ 160 K). Further, due to the energy conservation of the system the potential cluster energy in the stable state has to be smaller compared to that in the meta-stable

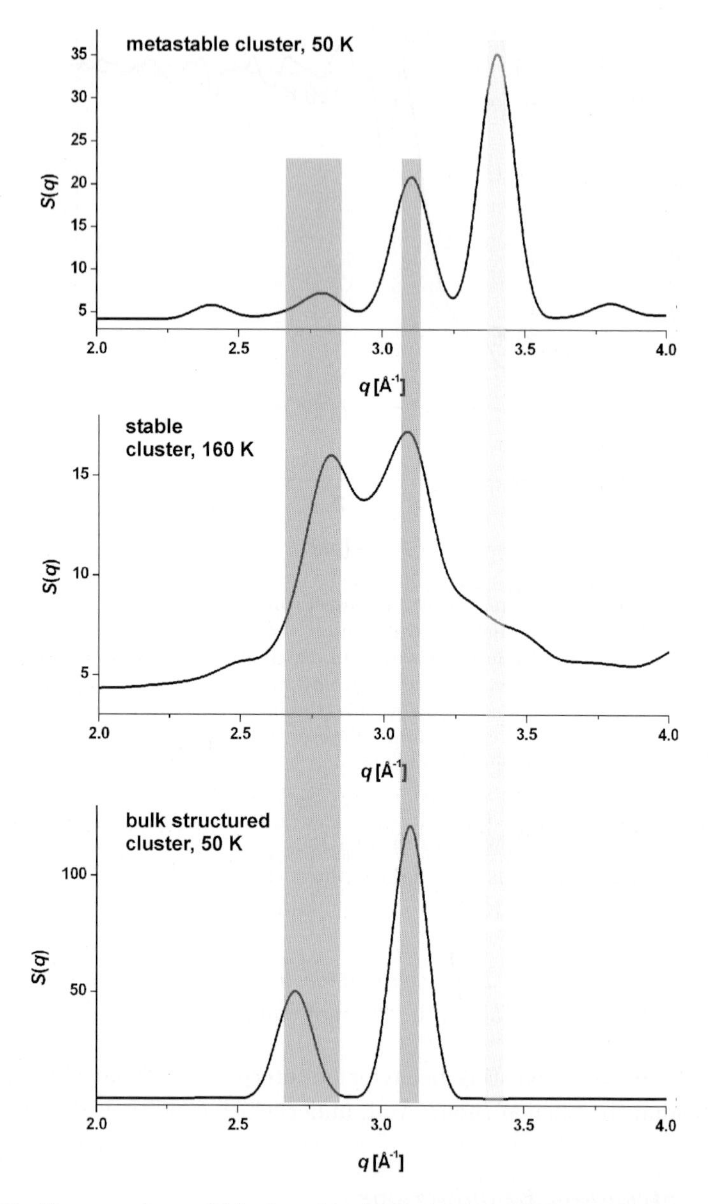

Fig. 5.11 Structure factor $S(q)$ of a cubic Al_{500} cluster in the meta-stable and stable state compared with $S(q)$ of a perfect fcc structured Al_{500} cluster with (001) surfaces: The dark grey bars indicate peaks which are common for all phases. They are typical for fcc structured aluminium. The light grey bar emphasizes the well pronounced peak at 3.4 Å^{-1} that appears only in the meta-stable state. This could be a hint for a superstructure.

state. Therefore, the structure of the meta-stable state should be different from that of the stable state, since the potential energy depends on the relative distances between the atoms.

For verification, the different cluster states have been investigated by means of the structure factor $S(q)$ (see Eqs. 4.29–4.32). Figure 5.11 illustrates the results for the meta-stable and stable state in addition to the solid bulk. As can be seen, the typical double peak of fcc structured crystals at about 2.7 $Å^{-1}$ and 3.1 $Å^{-1}$ appears clearly for all states (indicated by dark grey bars). A similar result has already been presented in Fig. 4.13 for solid aluminium at 800 K. But the meta-stable cluster state additionally exhibits a well pronounced peak at about 3.4 $Å^{-1}$ which is missing in the stable state as well as in the bulk (indicated by the light grey bar). This peak can be interpreted as a superstructure that has been formed within the meta-stable cluster due to a periodical particle shift.

A further conclusion can be drawn from the rather broad peaks in the structure factor of the stable cluster. These indicate that the structure of the stable cluster is no longer that of a perfect fcc crystal, but obviously, there are disordered components superimposed.

In Fig. 5.12, the various phases of an Al_{500} cluster are presented in front and cross-section view. The first row shows the initial cubic cluster before equilibration with the perfect fcc structure according to the bulk at 50 K.

The second row shows the cluster in the meta-stable state. Compared to the perfect crystal there are two obvious differences: the fcc structure is distorted at the edges and the volume of the cluster has decreased which leads to a higher density within the meta-stable state. Additional investigations using the pair correlation function $g(r)$ have shown that the positions of the peaks (which mainly result from neighboring atoms) have been shifted by about 0.3 Å compared to the bulk. The superstructure which has been recognizable from the structure factor $S(q)$ (Fig. 5.11), however, might result from this denser configuration.

The third and fourth row in Fig. 5.12 illustrate the cluster during the transition phase with temperatures of 100 K and 135 K. There is a structural change that starts from the faces of the cube. Some are bent inward and others belly out. This process continues step by step from the outside to inside until the stable state is reached after about 1 ps from the start of the transition (see Fig. 5.10).

The final structure of the stable state is presented in the last row of Fig. 5.12. The outer shape of the cluster has changed from a cube to a polyhedron. In addition, the hexagonal outline of the cross-section leads to

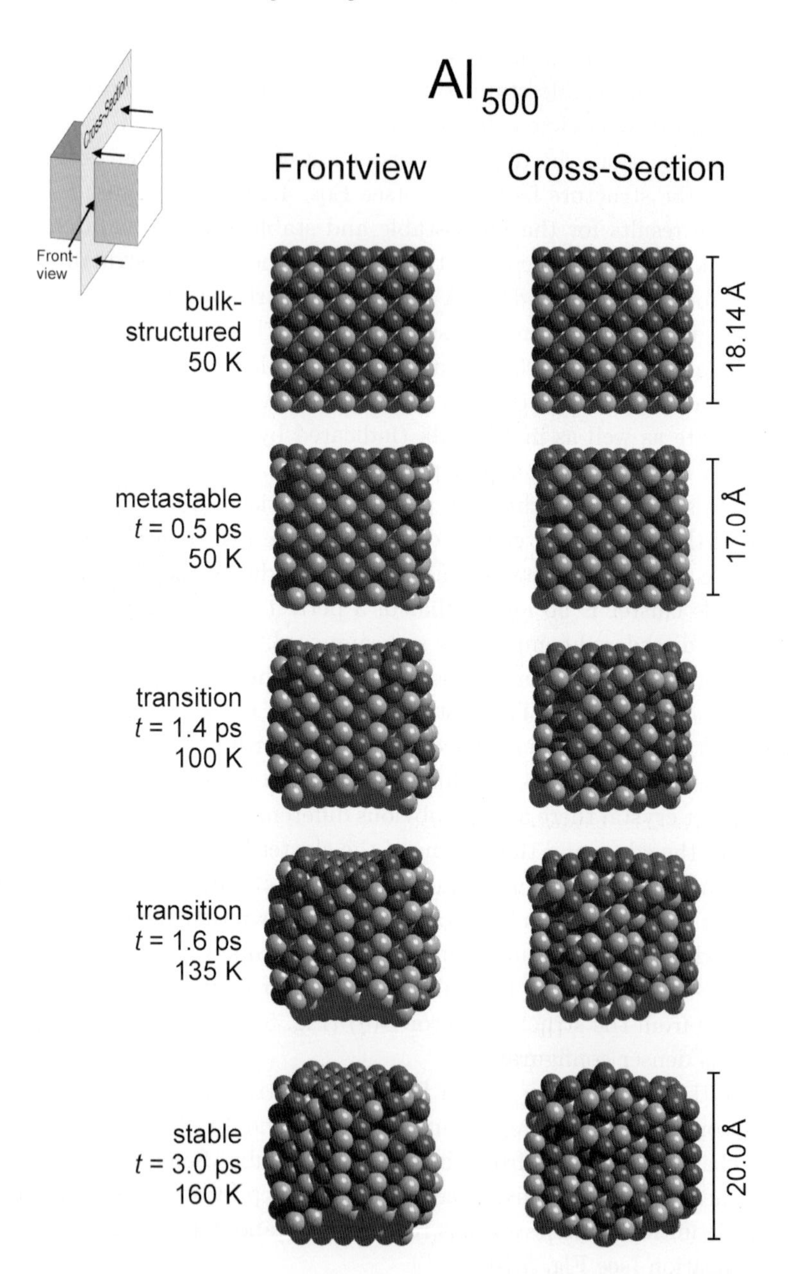

Fig. 5.12 Different phases of the Al_{500} cluster (the view directions are indicated by arrows in the upper left drawing as well as the orientation of the cross-section). The atoms of the initial layers are shaded different for a better understanding of the structural changes.

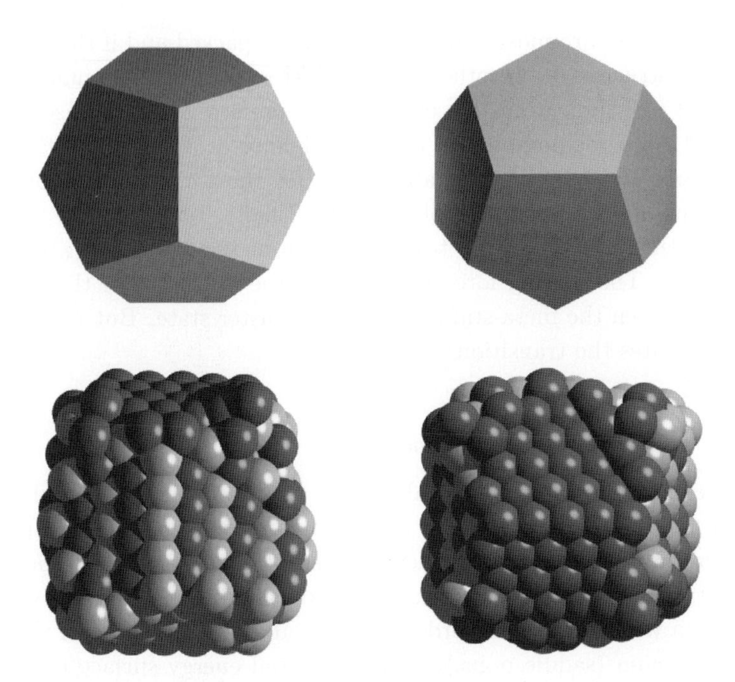

Fig. 5.13 Front and top view of the Al_{500} cluster in the stable state at 160 K. The atomic radii have been enlarged for a better visualization of the surface. In this way the resemblance to a dodecahedron (blue) can be seen quite clearly, even though the outer shape of the cluster is not perfect. There are dislocations near the edges that result from the multi-crystalline cluster composition.

the assumption that it could be a dodecahedron. And indeed, by looking from different directions onto the cluster — like in Fig. 5.13 — the supposition is confirmed. Further, the former (001) structured surfaces of the meta-stable cluster have been transformed into (111) structured faces of the dodecahedron. From the cross-section as well as from the outside, one can recognize dislocations. The stable cluster has no perfect crystalline structure. It is composed of various single crystals, i.e. it has a multi-crystalline structure.

From all these observations, we can draw the following conclusion: the stability of an Al_{500} cluster depends on its outer shape and structure, i.e. dodecahedral arranged (111) faces lead to a higher stability compared to (001) oriented cubic clusters.

This result fits nicely with [Ahlrichs, Elliott, 1999], where the stability of aluminium clusters with up to 153 atoms at 0 K have been studied using the *density functional method*. Here, the results have shown that clusters larger than Al_{55} are most stable if they are fcc packed and if their surfaces are (111) structured. In other words, clusters with a maximum number of nearest neighbor atoms are energetically preferred. In the case of Al_{500} clusters the average number of nearest neighbor atoms is 9.7 in the meta-stable state and 10.0 in the stable state (for the fcc structured bulk it is 12) which supports the result of [Ahlrichs, Elliott, 1999].

With this, the structural transformation of the initial Al_{500} cluster is reasonable. The driving force results from the difference of the potential energy between the meta-stable and stable cluster state. But which mechanism initiates the transition?

5.3.2 *Dynamics of the Al_{500} States*

When considering any meta-stable state the according potential energy surface $U(R_1, \cdots, R_N)$ (see Eq. 2.14) moves on a high level and prior to the decrease to the lower energy level of a stable state there has to be a more or less extended barrier which "keeps" the system at this higher level — at least for a while. In other words, a meta-stable state is given, if there is a local minimum (saddle point) on the potential energy surface neighbored by a even deeper minimum, and to transit from the higher to the lower level the system first has to overcome the local maximum that separates them.

In the case of the considered Al_{500} cluster this means that the meta-stable configuration remains in such a local minimum of the potential energy surface for about 1 ps. Then, without any external stimulation, it overcomes the existing energy barrier, leaves this state and transits into the stable configuration. To find the trigger of the transition we have to investigate the cluster dynamics by means of the *generalized phonon density of states* (PDOS).

Figure 5.14 shows the generalized PDOS $f(\nu)$ for the meta-stable and stable cluster state. Obviously, there is a significant difference between the frequency spectra: the distinctive peak at about 2 THz of the meta-stable cluster is not recognizable in the stable state (indicated by the light grey bar). Further, this peak is completely missing in the spectrum of the solid as well as in that of the liquid bulk (see Fig. 4.11).

In summary, only the meta-stable state is accompanied with the appearance of a pronounced low frequency oscillation mode at 2 THz. Taking into account that the velocity of sound in aluminium is 3130 m/s (at room temperature) [Wirdelius, 2000], then, at a rough estimate, transversal sound waves lead to resonance oscillations in the meta-stable Al_{500} cluster

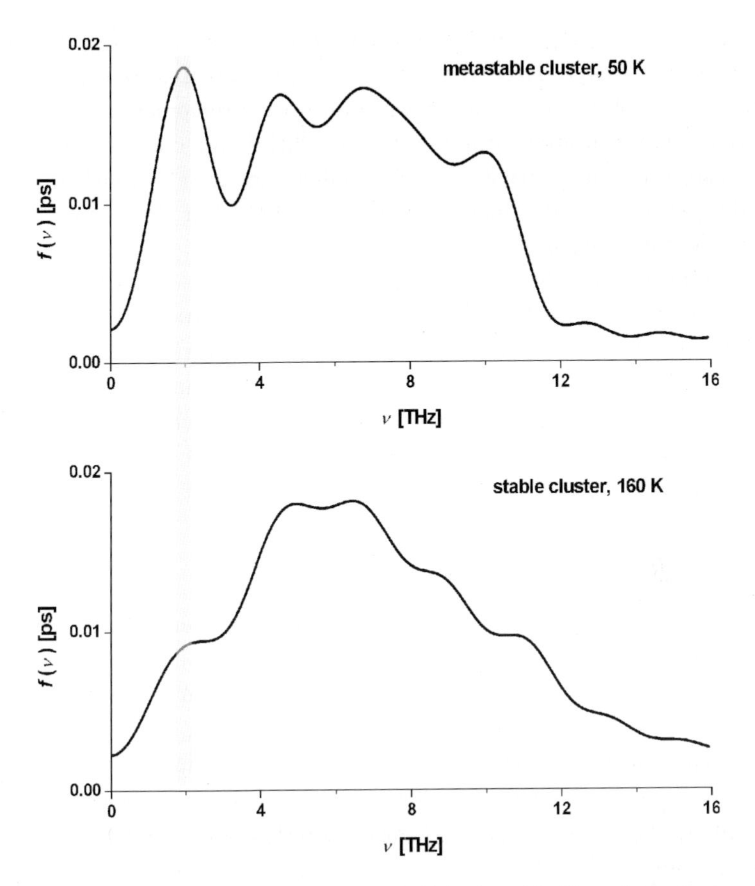

Fig. 5.14 Generalized phonon density of states $f(\nu)$ of a cubic Al_{500} cluster in the meta-stable and stable state: The well pronounced peak at 2 THz in the frequency spectrum (indicated by the light grey bar) of the meta-stable state is recognizable to some extent only in the stable state. This peak results from surface oscillations at the meta-stable cluster.

(side length 17 Å) with a frequency in the range of 1.84 THz. So the suspicion arises whether the 2 THz peak in the generalized PDOS of the meta-stable cluster could be due to surface oscillations.

To support this assumption digital movies have been prepared. Watching the sequence of the meta-stable period one can clearly see the oscillations. By a further look on the sequence of the equilibration phase it becomes obvious where the oscillations result from. As has already been mentioned in connection with Fig. 5.12, the outer dimensions of the meta-stable cluster are smaller compared to the initial bulk-structured cluster.

Therefore, during equilibration the whole cluster passes through a "shrinking process". Even if the temperature control steps are chosen to be rather small, there remains always an opposite momentum — comparable to the recoil after an implosion — which is directed outward. With increasing time the "shrinking and recoiling" behavior diminishes, but depending at which point the equilibration is stopped, it results in a more or less violent excitation of surface waves.

Now let us come back to the potential barrier that prevents the meta-stable cluster from an immediate transit to the stable state. Obviously, with the above mentioned pronounced surface oscillations it is only a question of time until a fitting configuration has established by chance which yields an energy high enough to overcome the barrier in the potential energy surface. In the following, the influence of several parameters on the meta-stable period are investigated.

5.3.3 *Influence of the Initial Conditions*

At first, the structural cluster transformation has been studied using different random velocity distributions for the MD models. There was no significant influence on the meta-stable period recognizable, the deviations from the above mentioned case have been about ± 0.3 ps. While in all studies the transitions into the stable states have led to dodecahedral structured clusters, the orientation of the final shapes have been different. Randomly, some looked like that illustrated in Fig. 5.13, the others like those presented in Fig. 5.15.

Fig. 5.15 Front and top view of an Al_{500} cluster: The initial velocity distribution has been chosen to be different from those of the cluster illustrated in Fig. 5.13. The comparison shows that the clusters differ in their spatial orientation. The dodecahedral axis are turned by 90°.

Other studies have been performed always using the same initial velocity distribution, but varying both equilibration parameters, the equilibration period and the number of temperature control steps (refer to section — *Isothermal Equilibration*).

Since the state transitions take place even under constant temperature conditions (*isothermal, canonical ensemble*) it is rather difficult to decide at which point to start the measurement for the meta-stable period. However, if equilibration is performed for times in the range of 0.5 to 2 ps, then the meta-stable period varies from about 0.5 to 2.5 ps. In addition, there is no systematic correlation with the number of equilibration steps, too.

In Fig. 5.16, typical temperature curves of cluster transitions are presented. As can be seen, all curves start at 50 K and end — after a more or less extended meta-stable period and after quite different transition behaviors — in the temperature range of 150 to 160 K. The meta-stable period is not definitely correlated to the kind of equilibration.

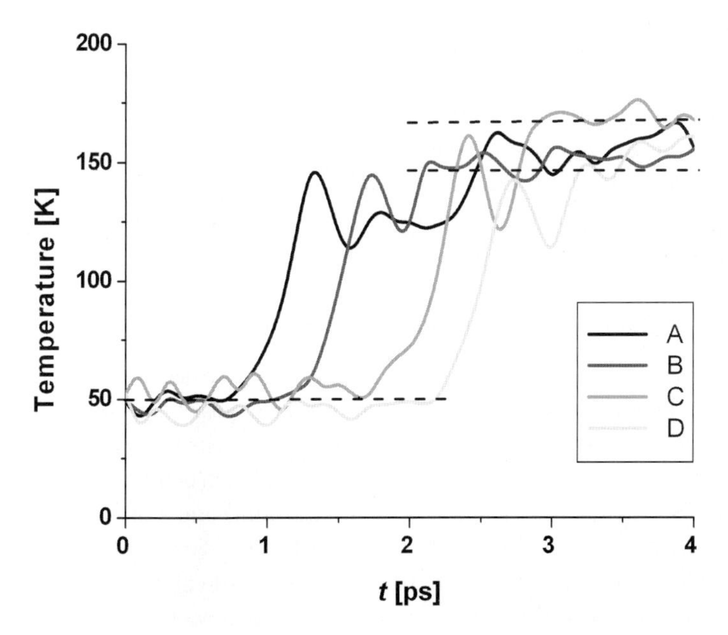

Fig. 5.16 Temperature transition curves from the meta-stable to stable state of the same Al_{500} cluster with different prior equilibration phases: (A) $T_E = 2$ ps, $n_E = 20$, (B) $T_{E1} = 0.5$ ps, $n_{E1} = 5$ followed by $T_{E2} = 1.5$ ps, $n_{E2} = 15$, (C) $T_E = 1$ ps, $n_E = 10$, (D) $T_E = 1.1$ ps, $n_E = 10$. (T_E: equilibration period, n_E: number of time steps after which temperature control is performed.)

5.3.4 *Influence of the Initial Temperature*

When raising the initial cluster temperature, well extended meta-stable phases — the period varies around 1.5 ps — can be observed for at least up to 200 K. But for initial temperatures of about 250 K and above, the transition takes place nearly immediately (about 0.3 ps) after the end of equilibration with a preceding drop of the cluster temperature before its raise to a significantly higher stable level. It seems, that due to higher initial temperatures the potential energy barrier can be overcome easily and, therefore, meta-stable states cannot establish.

On the other hand, there is an additional interesting effect observed, for example, for an initial temperature of 350 K. While most transitions — even at higher temperatures — result in dodecahedral shaped clusters as illustrated in Figs. 5.13 and 5.15, there are cases deviating from this behavior. Figures 5.17(b), (c) and (d) show the resulting structure from different viewing points. At 430 K, this stable configuration has the shape of a cuboid where four surfaces are (111) and the other two are (001) structured.

Beside this, there are still other cases in which the structural transformation takes place only partly. This has been observed, for instance, for an initial temperature of 450 K which ended at 530 K in the stable state. Here, the resulting clusters consist of dodecahedral as well as of rectangular shaped parts. In these cases (001) and (111) surfaces are mixed by chance. Mainly, such partial transformations are also observed with larger clusters at low temperatures as will be discussed later on.

5.3.5 *Influence of the Crystalline Structure*

Depending on the alignment of a cube within the fcc bulk structure the resulting cluster is composed of a pile of layers with a specific orientation. In the case of the above mentioned studies the orientation of these layers has been (001) with an alignment parallel and perpendicular to the unit cell. If the alignment of a cube is rotated around one axis by 45° (refer to Fig. 3.2), the resulting initial cluster looks like illustrated at the left side of Fig. 5.18(a). It consists of four (011) and two (001) structured faces.

As can be seen, the cluster corresponds not exactly to a close packed configuration. That's why such clusters are not stable. They transform into another structure immediately after the end of equilibration without any hint of a growing meta-stable state. The resulting structure (Fig. 5.18(a), right side) completely consists of (001) surfaces with 45° alignment.

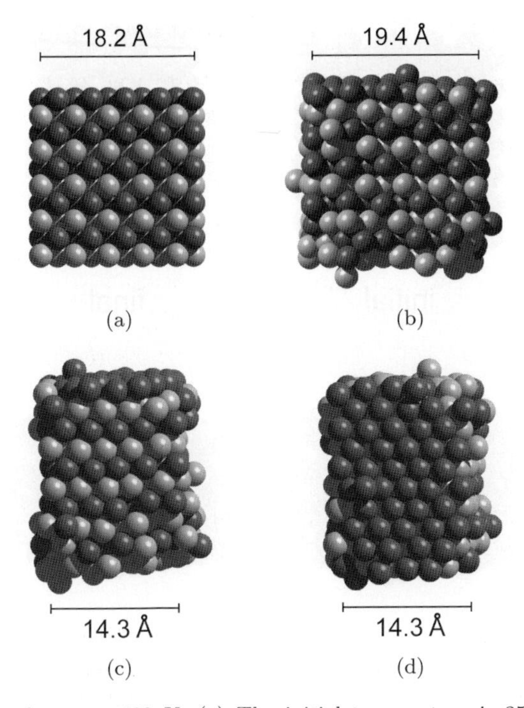

Fig. 5.17 Al_{500} cluster at 430 K: (a) The initial temperature is 350 K, the cluster shape is cubic with (001) surfaces. After equilibration for 1 ps, the cluster transforms immediately into a stable configuration: (b) Front view, (c) side view, (d) top view. Here the shape of the stable configuration corresponds to a cuboid with (001) structured front and back sides and four (111) structured faces.

Figure 5.18(b) presents an Al_{474} cluster consisting of (111) layers. During the equilibration process a slight increase in density is observed (like the above mentioned shrinking of Al_{500} clusters), but no structural transformation is taking place.

Without any question, the cluster state is strongly influenced by the crystalline structure. Depending on the orientation of the surface layers the considered cubic aluminium clusters are either totally stable ((111) surfaces and (001) surfaces with a 45° alignment) or unstable ((011) surfaces). Only in the case of (001) structured surfaces with 90° alignment, meta-stable states have been detected under the above mentioned conditions.

5.3.6 *Influence of the Outer Shape and Cluster Size*

The prior investigations lead to the reasonable assumption that the outer shape has a lasting influence on the cluster state. Examples already have

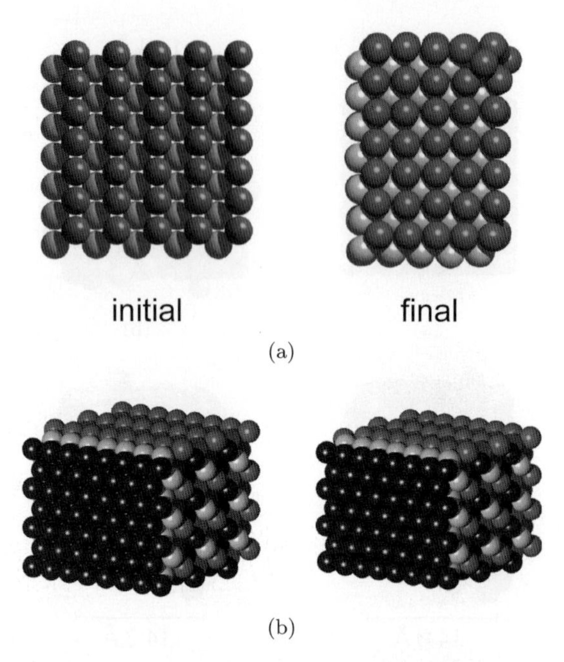

Fig. 5.18 The initial and final structures of Al_{490} and Al_{474} clusters: (a) The initial Al_{490} cluster is assembled with (011) layers and, therefore, its surface consists of four (011) and two (001) structured faces. Since that is not a close packed configuration, the cluster immediately transforms into a denser state with (001) oriented planes and surfaces only. (b) The Al_{474} cluster is composed of (111) layers. This is a very stable configuration and its structure does not change further.

been given in Fig. 4.1. Depending on the kind and complexity of the cluster shape the structures should be more or less stable. With respect to the previous studies, cluster shapes that primarily lead to (111) surfaces should be most stable. Therefore, spherically shaped nanostructures show no tendency of transformation and rectangular shaped clusters are seemingly the least stable ones. The stability of all other cluster shapes varies between those two opposites. Since it makes no sense to sample through all possible cluster shapes, the following MD studies focus on the influence of the size of cubical clusters on their stability.

It has been already discussed that the larger the cluster the smaller the influence of the surface, since with increasing cluster size the ratio of surface to core atom numbers decreases. So, this should significantly influence the meta-stable states. Following the examples of the Al_{500} clusters, the appearance of meta-stable states and structural transformations have been investigated for initially cubical clusters in the range of less than 256 and

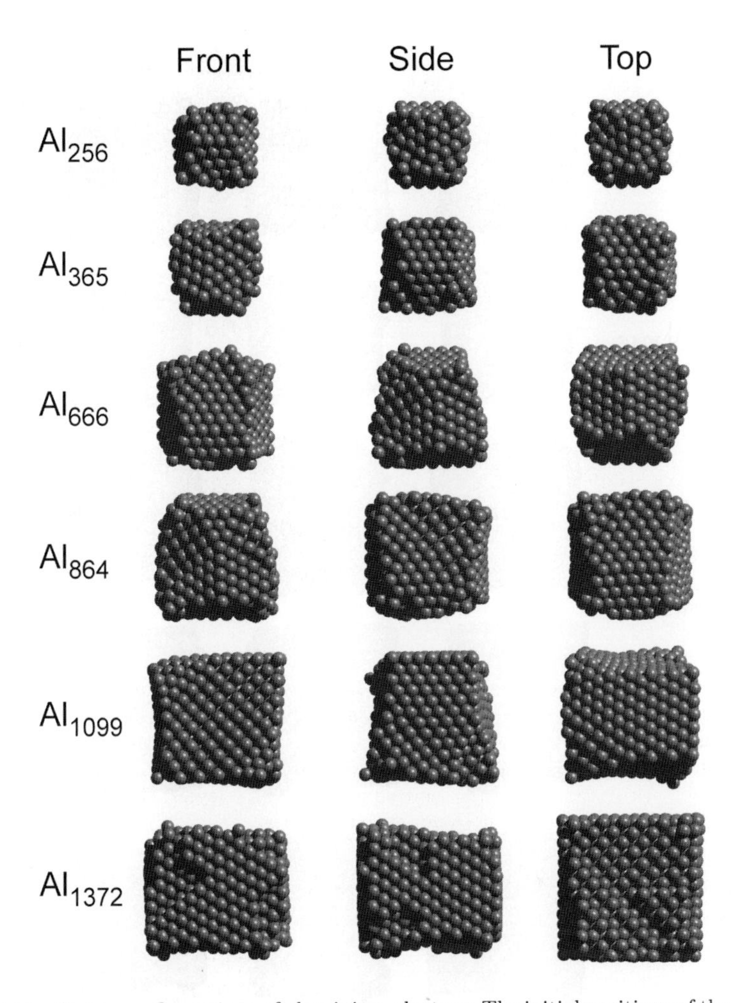

Fig. 5.19 Stable configurations of aluminium clusters: The initial positions of the atoms have been chosen to be those of the fcc bulk structure of aluminium and the outer shapes have been cubic with (001) oriented surfaces. The Schommers potential with a cut-off radius of 7.35 Å has been used to describe the inter-atomic interactions. After the end of an equilibration phase all clusters have entered a meta-stable state at 40–50 K prior to transitions into the final stable configurations that are illustrated from different viewing directions.

up to 4000 aluminium atoms. Depending on the equilibration procedure, the initial temperatures have been 40–50 K.

Figures 5.19 and 5.20 give an overview of the finally stable configurations that have been preceded by meta-stable states. Though using a lot of different models, meta-stable states could not be detected for Al-clusters

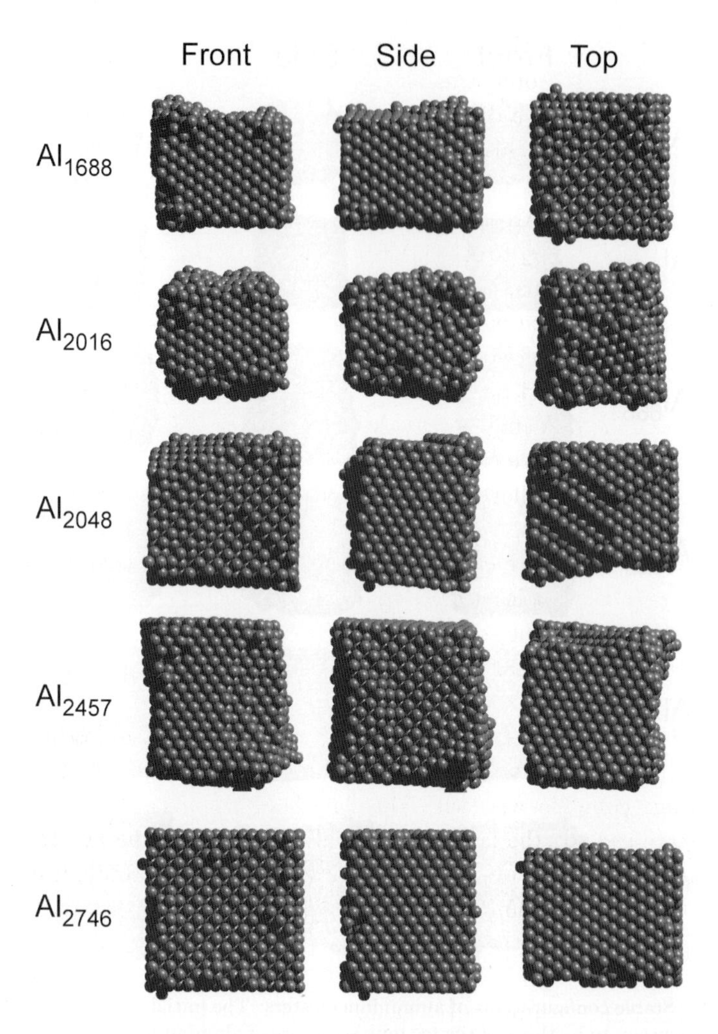

Fig. 5.20 Stable configurations of aluminium clusters (continuation from Fig. 5.19).

smaller than 256 atoms, here transitions into stable states took place immediately.

On the other hand, clusters larger than 2748 atoms, e.g. Al_{2916} or Al_{4000}, have been proven stable (at least for 150 ps). Such large clusters remain in the "meta-stable" state without transition.

All other clusters from Al_{256} to Al_{2746} show more or less extended meta-stable periods. A systematic correlation between meta-stable period and cluster size is nearly impossible, since the meta-stable states depend on the initial conditions and, therefore, one never can be sure whether or not there are still other initial configurations that lead to more extended meta-stable periods. Nevertheless, a division into three categories seems to be feasible.

The first group of clusters (Al_{256}, Al_{365}, Al_{666}) remain in meta-stable states for a maximum of about 2–3 ps. The meta-stable period of the second group (Al_{864}, Al_{1099}) reaches values of 7 ps at least. And clusters of the third group (Al_{1372} to Al_{2746}) have been observed to be in meta-stable states for at least 20 ps.

Again, it has to be emphasized that this division bases on the result of "only" fifty different MD models, while for a precise analysis, billions of MD calculations would be necessary. Even though the given division may be improved, the tendency is reasonable: the larger the clusters, the higher the probability for more extended meta-stable periods.

There is still another reason that supports the division of the clusters into three groups. The surfaces of the clusters of the first group (Al_{256}, Al_{365}, Al_{666}) are exclusively (111) structured and the outer shape is dodecahedral (or at least close to a dodecahedron).

The structures of the second group (Al_{864}, Al_{1099}) look like a mixture of dodecahedral and cubical parts. Their surfaces show (111) as well as (001) structured areas.

And the third group (Al_{1372} to Al_{2746}) consists of more or less perfect cubically shaped clusters with four (111) and two (001) oriented faces. These observations coincide well with the length of the according meta-stable cluster state period: the larger the initial cluster cubes the less the probability for a transformation into an energetically favorable configuration (the free energy shows a minimum) with overall (111) oriented surfaces and the longer, the possible meta-stable period.

The only exception is the Al_{2016} cluster for which intentionally a stable configuration with overall (111) surfaces has been selected to be shown within Fig. 5.20. This should emphasize the probability aspect of the considered cluster transformations. The large clusters of the third group most often transform into a cuboid-like form, but in some cases — depending on the initial configuration — they take quite a different shape.

This random behavior is demonstrated in Fig. 5.21 for initially cubic Al_{2048} clusters. At the beginning the atomic positions have been identical, but for each cluster different velocity distributions have been used (either by the random generator or by different equilibration parameters) in order to create dissimilar equilibrated configurations. As can be clearly seen, depending on randomly selected initial configurations the shape as well as

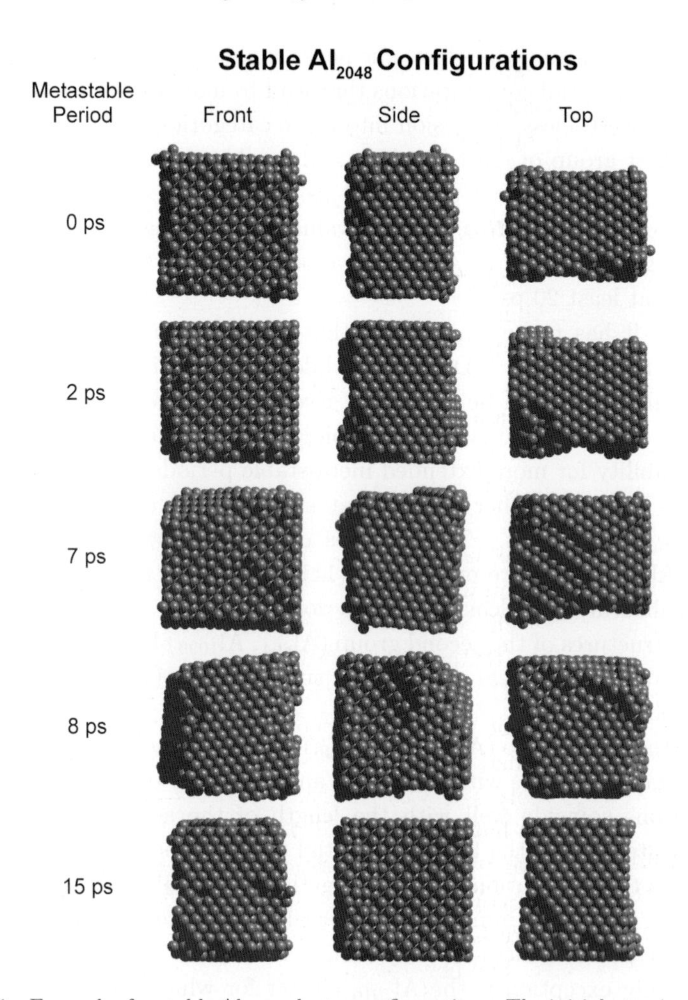

Fig. 5.21 Examples for stable Al_{2048} cluster configurations: The initial atomic positions are identical, but the velocity distributions have been chosen to be different. The MD parameters match those of Figs. 5.19 and 5.20.

the poly-crystalline arrangement of the final stable clusters diverge significantly. The duration of the meta-stable states too varies between 15 ps and immediate transition (0 ps) depending on the initial cluster configuration.

In summary, the MD studies have shown that the occurrence of meta-stable states is determined by the cluster size. Further, the transition into a stable configuration is a random event. Depending on the initial conditions the meta-stable periods may vary within a broad range — a prediction is not possible. But the mean duration of the meta-stable states seems to be dependent on the cluster size, since the MD results allow for a division of cubic aluminium clusters into three size categories which, in addition, are correlated to the most probable outer shapes.

5.3.7 *Influence of the Interaction Potential (Material)*

With respect to the oscillating form of the temperature dependent *Schommers* potential for aluminium (see Fig. 2.2), the question arises whether the investigated structural transformation of nano-clusters is restricted to this specific interaction potential or material, respectively. Though a lot of important parameters and properties have been described correctly using the *Schommers* potential, there might be, nevertheless, small inaccuracies that artificially lead to the described structural transformation effect for aluminium.

Therefore, extensive investigations have been performed to find a similar behavior with krypton clusters using the *Barker* potential (Table 2.4, Fig. 2.1). Since this effective pair potential is one of the most accurate ones and, in addition, temperature independent, it may be considered as model potential for MD calculations. If meta-stable states and structure transformations could be found with krypton clusters using the *Barker* potential, this certainly had to be regarded as proof for their occurrence with aluminium and possibly with other materials, too.

The investigations have been started for cubical Kr_{500} clusters at 10–50 K without positive result: the clusters have been extremely stable. Smaller cubical clusters of different sizes too have led to no result, although the according structure factors contain additional peaks which cannot be observed in the bulk. Further, the distinct low frequency peak within the generalized phonon density of state function — which has been typical for the meta-stable aluminium clusters — is only weakly pronounced with these small cubical krypton clusters.

But finally, cuboidal or, more precise, bar shaped clusters have shown the same behavior like that of cubical aluminium clusters. Figure 5.22 illustrates a representative example of the meta-stable state and the transition into the stable configuration of a Kr_{313} cluster.

As can be suspected from Fig. 5.22, the meta-stable state is accompanied by cluster oscillations. For some clusters, this has been confirmed by the appearance of well pronounced low frequency peaks (0.3–1.2 THz) in the generalized phonon density of state function which coincides with the observations made with aluminium clusters. Due to the weaker Kr–Kr

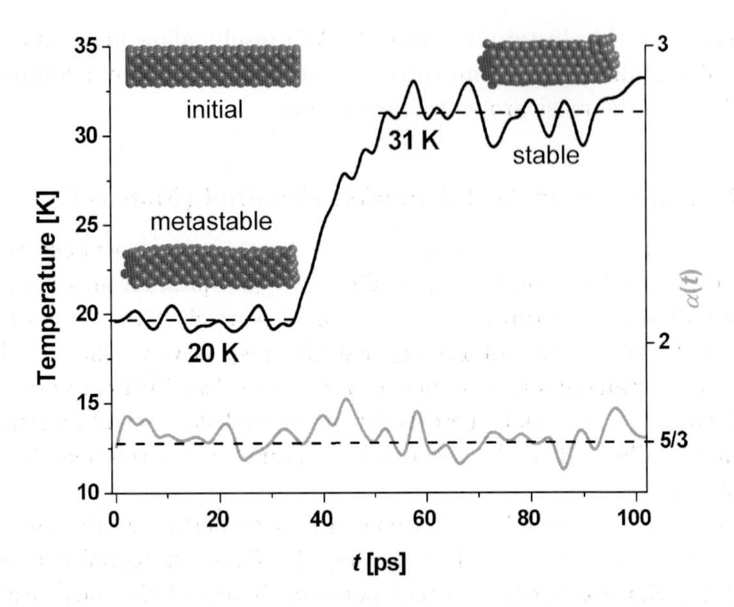

Fig. 5.22 Temporal development of the temperature and α-function for a bar shaped Kr_{313} cluster with (001) surfaces (the initial, meta-stable and stable configurations are illustrated within the graph): Initially, the cluster remains in thermal equilibrium at about 20 K for 33 ps ($\alpha = 5/3$ indicates Maxwellian distribution). Then, during a transition phase the cluster temperature increases up to about 31 K. The MD calculations have been performed using the Barker potential with a cut-off radius of 10 Å. Equilibration has been performed for 2 ps.

bonds the increase in temperature during the transition phase is less and the meta-stable period is longer compared to aluminium clusters.

The structural transformation is based on a contraction of the krypton cluster which finally results in (111) oriented surfaces, again, comparable to Al-clusters.

In summary, the MD studies with krypton clusters give rise to the assumption that meta-stable states and structural transformations are not restricted to certain elements only. Within a more or less narrow range of parameters these phenomena should be observable for many (maybe all) materials with *van-der Waals* or metallic bonds.

5.3.8 *Conclusions*

Regardless of the lattice structure of the bulk, small nano-clusters try to form such configurations for which the potential cluster energy takes a minimum value. Since these configurations coincide with a maximum number

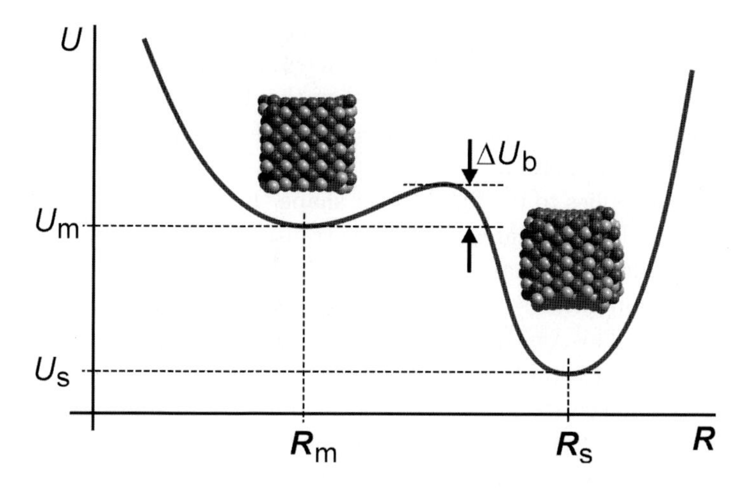

Fig. 5.23 Schematic drawing of a two-dimensional projection of the potential cluster energy surface $U(\mathbf{R})$: If an Al_{500} cluster takes a configuration near the point $\mathbf{R}_m$ of the configuration space $\mathbf{R} = [\mathbf{R}_1, \cdots, \mathbf{R}_{500}]$, it is locked up within the surroundings of the local minimum (saddle point) of the potential cluster energy U. Due to energy fluctuations — caused by the statistical temperature movements of the atoms, enhanced by surface or body oscillations — after some time the cluster configuration by chance may take a point at which the energy barrier $\triangle U_b$ has been overcome. At this point, the cluster transforms into the configuration $\mathbf{R}_s$ — the stable state.

of nearest neighbor atoms, the cluster structure often deviates significantly from that of the bulk, i.e. structurally disturbed clusters are the rule.

However, depending on size, shape and material, local minima (saddle points) of different depth may appear in the potential cluster energy surface. Under certain conditions, a cluster configuration can be "locked up" in a local minimum that is surrounded by even deeper minima (see Fig. 5.23). In this case the cluster takes a meta-stable state.

Due to the temperature fluctuations of the atoms, sometimes enhanced by oscillations, the cluster configuration permanently fluctuates around the configuration point $\mathbf{R}_m$. Depending on the height of the energy barrier $\triangle U_b$ and the distance between the minima, it takes some time — the meta-stable period — after which a configuration point is reached by chance that enables the cluster to overcome the barrier. After that, the cluster transforms into the stable configuration $\mathbf{R}_s$ and the decrease of potential energy goes with an increase of temperature.

Using this picture, a plausible interpretation of the results from the previous MD studies is possible:

- The initial conditions of the MD calculation determine the configuration point. It can be located near the barrier, then the meta-stable period is very short. Or it may be far away, then the duration of the meta-stable state is rather long. This is the explanation for the probability aspect of the meta-stable period.
- The same applies to the final cluster shape. It is a matter of chance into which neighboring local minimum the transition takes place. Clearly, the initial configuration has an impact on the final shape.
- At high temperatures, the energy fluctuations are larger than the potential barrier. This, on the other hand, prevents the cluster from remaining within the meta-stable state.
- For very small clusters there are no local minima available in the potential energy surface that could provide meta-stable states. The same applies to cluster configurations with (011) surfaces.
- Cluster configurations with (111) surfaces are already near a deep minimum. In this case, a transition cannot take place.
- For very large clusters the barrier between the local minima is too high, and in addition, the temperature fluctuations are getting smaller with increasing particle numbers. Then the cluster remains in the "meta-stable" state forever (at least without external influence).

The last point is very interesting. The MD studies with cubic aluminium clusters containing more than 2746 atoms have not led to the observation of state transitions. However, the above mentioned conclusion might be proofed, if one of these large aluminium clusters could be stimulated to perform a structure transformation. This is the topic of the following section.

5.4 Stimulated Nano-Cluster Transformations

It has already been discussed that large cubic aluminium clusters with (001) oriented surfaces build a rather stable phase while smaller cubes with the same face orientation show more or less extended meta-stable states which finally transit into stable configurations with mainly (111) structured surfaces.

As an example, without external influence an Al_{2916} cluster (upper most picture of Fig. 5.24) does never leave such a "stable" state. This has been proofed for several initial conditions and periods of up to 150 ps. Since the considered MD models mirror ideal situations only, where no interactions with the environment take place (absolute vacuum), the question arises,

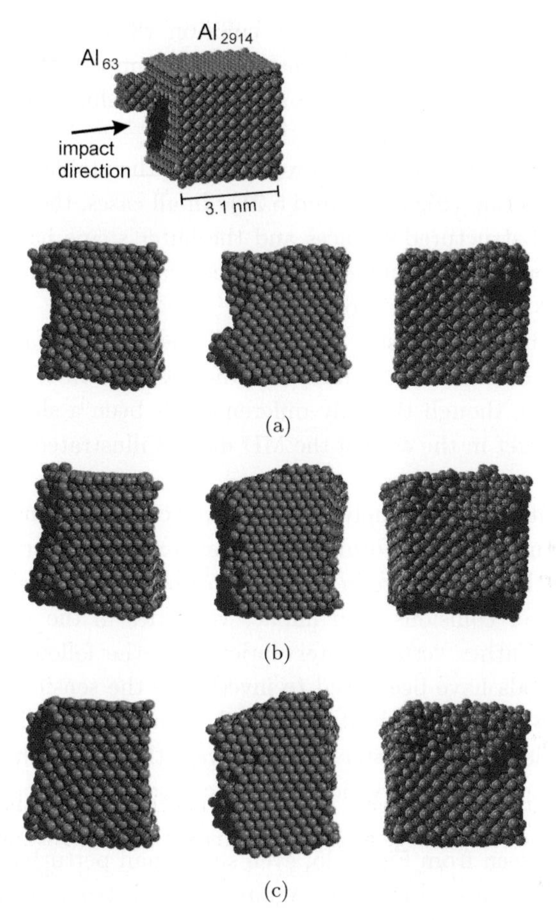

Fig. 5.24 The upper most picture shows a "pseudo stable" state of an Al_{2916} cluster (the layers are colored red and green for a more instructive visualization of the structure) that can be stimulated to transit into finally stable states by a collision with another cluster (in this case an Al_{63} cube which is colored blue). The final structure strongly depends on the impact parameters. Several cases are illustrated, each from three different viewing angles (front, top and impact direction), about 3 ps after the cluster collision: (a) The impact velocity of the Al_{63} cluster is 1000 m/s. (b) The impact velocity is 2000 m/s with the same impact target as in (b). (c) The impact velocity is 2000 m/s but the target has been slightly shifted by about 0.2 nm to the upper left side.

whether an external stimulation can trigger a structural transformation in the same way as has been observed for smaller clusters without external influence.

An obvious method to induce a perturbation of the Al_{2916} cluster configuration is to use another cluster for a collision. Figures 5.24(a)–(c) show three final structures 3 ps after the collision with an Al_{63} cluster with different impact velocities and slightly different impact targets. It can be seen quite clearly that in all cases structural transformations have taken place.

The final configurations are comparable to the stable cluster states of the previous section (Figs. 5.20 and 5.21). In all cases, there are four (111) and two (001) structured surfaces and the outer shape has changed from a cube to more or less cuboid forms. Dislocations and multi-crystalline structures are recognizable too.

Further, the probabilistic character of the transformation — due to different initial conditions — mirrors in the observation of clearly different final structures, though the only difference has been a slight deviation of the impact target in the case of the MD models illustrated in Figs. 5.24(b) and (c).

As a result, even seemingly stable cluster configurations may be stimulated to structural transformations by an external intervention, i.e. in reality these clusters too are in "pseudo" or meta-stable states.

Since cluster collisions with impact velocities in the range of several 1000 m/s are rather "crude" interventions, for the following MD models "softer" methods have been used to investigate the sensitivity. Therefore, an Al_{63} cluster has been heated up and positioned within reach of the "pseudo stable" Al_{2916} cluster. In this way, the larger cluster is heated locally at the contact spot which may be considered as local perturbation of the surface structure.

As can be seen from Fig. 5.25, even such small perturbations are sufficient to trigger phase transitions. Again, the final structures have the form of cuboids with four (111) and two (001) oriented surfaces. Here too, the totally different final structures [Figs. 5.25(a) and (b)] — due to the just slightly different temperatures of the Al_{63} cluster only — demonstrate the random nature of the transformation process.

The investigations of the previous sections have shown that the cluster stability improves with increasing cluster size. Therefore, it has to be suspected that there is a limit above which the configurations are globally stable, independent of cluster collisions or other perturbations. Of course, this limit still has to be within the nanometer scale, because macroscopic single crystals are definitely stable.

However, large-scale MD studies could be performed to find such limits. Since this would exceed the frame of the current monograph, it will be left for future investigations.

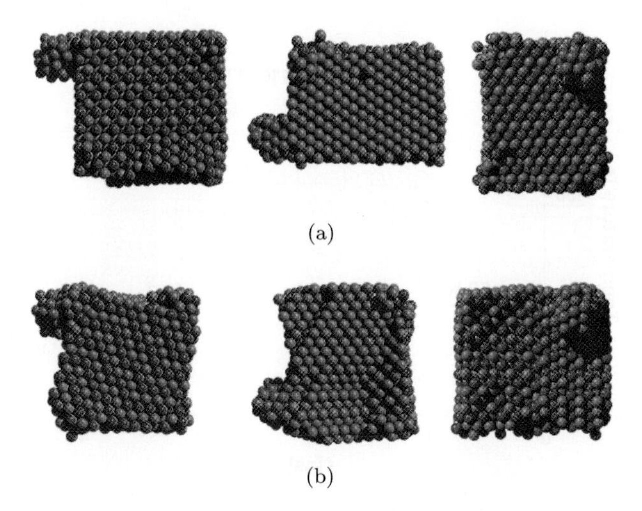

Fig. 5.25 Initially, a heated Al_{63} cluster (blue) has been positioned within reach of an Al_{2916} cluster (the same as in Fig. 5.24) at a temperature of 40 K. After contact has been established, the surface structure of the large cluster is perturbed locally. This initiates a structure transformation. Depending on the initial conditions — here the temperature T_i of the Al_{63} cluster — the final shapes and structures of the Al_{2916} cluster deviate significantly from each other. (a) $T_i = 350$ K, (b) $T_i = 400$ K. (The viewing angles and colors have been chosen to be the same as those in Fig. 5.24.)

5.5 Analogy Considerations

After all these investigations of the influence of certain parameters on meta-stable cluster states it might be helpful to build some vivid connections.

Looking through the previous MD results, a certain analogy can be drawn between the meta-stable cluster states and the excited states of atoms (see Fig. 5.26):

- The potential energies of the meta-stable (excited) states are larger than those of the stable (ground) state.
- After a certain period the clusters (atoms) transit from their meta-stable (excited) states into the stable (ground) state without external influence, the atoms by emitting photons, the clusters by increasing their temperature.
- The period (lifetime) of the meta-stable (excited) states is not constant, but probability distributed. More or less, all intervals are possible, but the values with maximum probability are most likely to appear.
- The transition from meta-stable (excited) to the stable (ground) state can be triggered by an appropriate stimulation, the electrons by photons

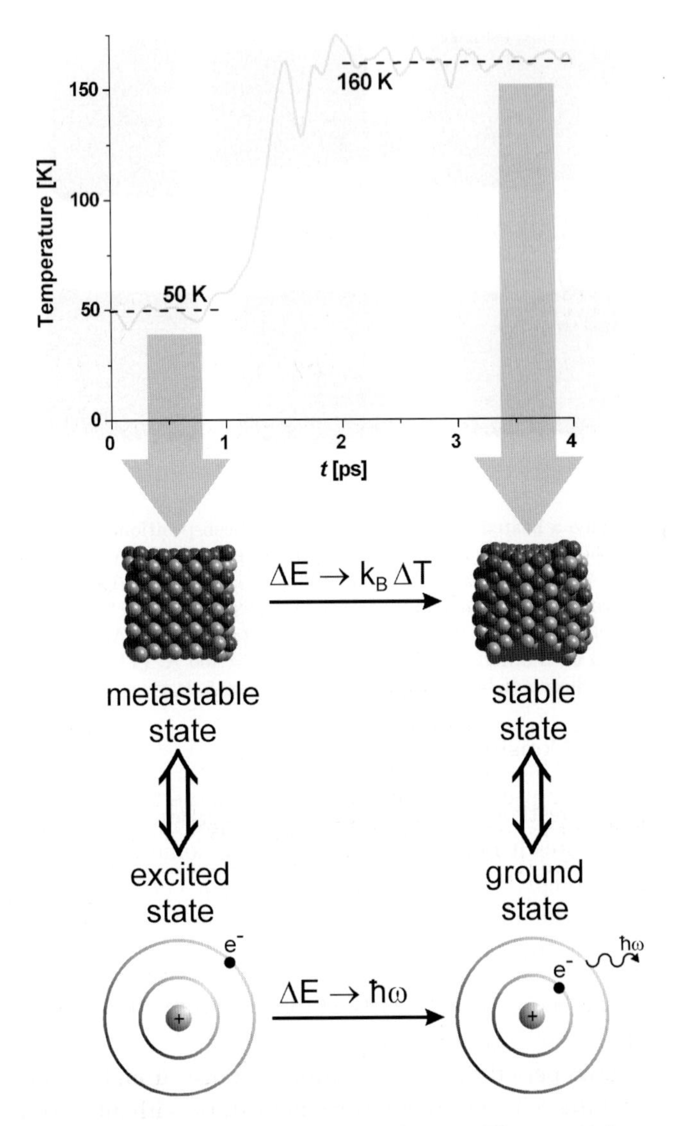

Fig. 5.26 The analogy between nano-clusters and atoms: The potential energy of both, the meta-stable cluster state and the excited state of atoms, is on a higher level than the according stable state and ground state, respectively. After a certain time (meta-stable period for clusters and lifetime for excited atoms, respectively) transitions into the lower energy states take place without external influence. The atoms emit photons, the clusters increase their temperature. (More details are given in the text.) Further, the transition of meta-stable to stable cluster states as well as excited atoms can be triggered by an appropriate stimulation. The values for both, the meta-stable period and the lifetime, are probability distributed.

of certain wave lengths, the clusters by a local perturbation of the surface.

While the lifetime of the atomic excited states lasts for about 10 ns, the magnitude of the meta-stable periods of aluminium cluster configurations is 10 ps. Compared to the lifetime of excited atoms this seems to be rather short. But with respect to the time interval, a sound wave needs to propagate across a nano-cluster (for an Al_{2048} this is about 0.5 ps) which it is a considerably long period.

However, there is at least one point for which the analogy considerations with the atomic states fails: for nano-clusters there is a huge number of stable configurations (for an example, see Fig. 5.21), but atoms have only one ground state. This fact leads to another analogy, the *bifurcation phenomenon*.

5.6 The Bifurcation Phenomenon at the Nanometer Scale

In connection with meta-stable states of small clusters as well as with "pseudo stable" states of larger clusters the following effect (which has already been discussed in the previous sections) is of particular interest.

When a cluster makes the transition from a meta-stable or pseudo stable state to a stable state, there are several possibilities. In other words, the local minimum (saddle point) in the potential energy surface correlated to the meta-stable or pseudo stable state is surrounded by numerous deeper minima which correspond to the stable states. Which of these stable states the cluster will finally take cannot be predicted in principle. Due to the unavoidable energy fluctuations of nanosystems, it is completely a matter of chance.

That is, a cluster state transition is a *bifurcation* in the sense of *chaos theory* (see Fig. 5.27). Metaphorically speaking, at the *bifurcation point* nature plays dice to decide on which of the various branches (stable states) the cluster will finally rest.

The numerous MD studies presented in the previous sections have shown that stable cluster configurations may differ in both the inner structure as well as the outer shape. In general, the structures show multi-crystalline compositions, i.e. there are grain boundaries, dislocations and other lattice defects. And finally, the outer shape strongly depends on the arrangement of these lattice defects.

With respect to the structure transformations of nano-clusters the bifurcation phenomenon has no counterpart, neither at the micrometer scale nor

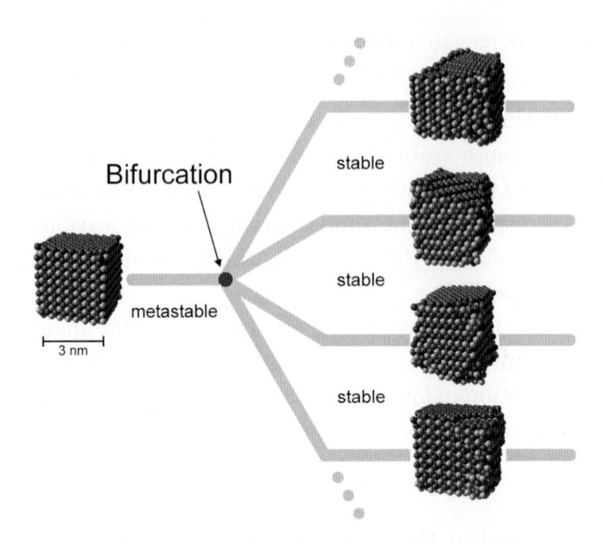

Fig. 5.27 A vivid illustration of the bifurcation phenomenon: The stable state into which a meta-stable cluster will finally transit cannot be predicted. Metaphorically speaking, at the bifurcation point nature plays dice to decide on which of the various branches the cluster will finally rest.

within the macroscopic bounds. If this would not be the case, any object which was in a meta-stable state could transform randomly into another shape, e.g. after some time a meta-stable beer can could transform into a cookie box or into a fork or into something completely different. But, as we know, in reality the only transitions of macroscopic objects result from interactions with the environment — signs of wear — which finally lead to their destruction.

For nano-clusters the situation is completely different: the transition at the *bifurcation point* is *constructive* (Fig. 5.27), since the cluster is not to be destroyed, but transforms into a new shape that is clearly more complex than the initial meta-stable cube.

This behavior of single clusters, resulting from the mutual influence of temperature and particle interaction, mirrors a certain kind of *independent creativity* which is an inherent characteristic of the nanosystem itself.

5.7 Analogies to Biology

Although aluminium clusters represent *inorganic* structures there are several analogies to biological systems [Rieth, Schommers, 2002]:

- A nanosystem may be creative, i.e. it can take individual decisions which cannot be influenced from the outside. This is an interpretation of the bifurcation phenomenon.
- Nanosystems are able to transform spontaneously, either stimulated by a random perturbation or even without any external influence. Such a behavior is also known from biological systems: spontaneous transformations are typical phenomena within the field of embryology [Beloussov, 2000, 2002].
- The state transition is accompanied by a structural transformation where new shapes are created. Also within the field of embryology new forms appear which have not existed before.
- Simple shaped meta-stable nano-cluster transform into more complex configurations. This behavior too is typical for biological systems — it is a feature of the evolution.

5.8 Final Considerations

As has already been emphasized, the key element in *molecular dynamics* is the *interaction potential*. Even if a considerable amount of work is spent to establish a potential function that is able to reproduce all known experimental results within acceptable accuracy, there always remains an uncertainty, since small deviations in the potential may lead to substantial differences in the MD results. But now, of all things, the use of different cut-off radii for the potential function — which is a necessary and common technique of MD calculations — leads to such deviations.

For asymptotically vanishing functions like that for krypton there is no problem: to be on the sure side one just has to increase the cut-off distance to a value where the potential is negligible. But for metallic bonds the according pseudo potentials show rather far-reaching oscillations that are only negligible for long distances. On the one hand, to consider all these oscillations would lead to ineffective MD algorithms. On the other hand, it is questionable at all whether the real nuclear interactions are this far-reaching. In the case of the *Schommers* potential for aluminium, the use of a cut-off radius in the range of the six nearest atomic neighbor distance (7.35 Å at 50 K) has delivered all the important characteristics and effects in good agreement with experiments.

Another point to discuss is the practical relevance of the observed cluster state transitions. Since standard mass production techniques for

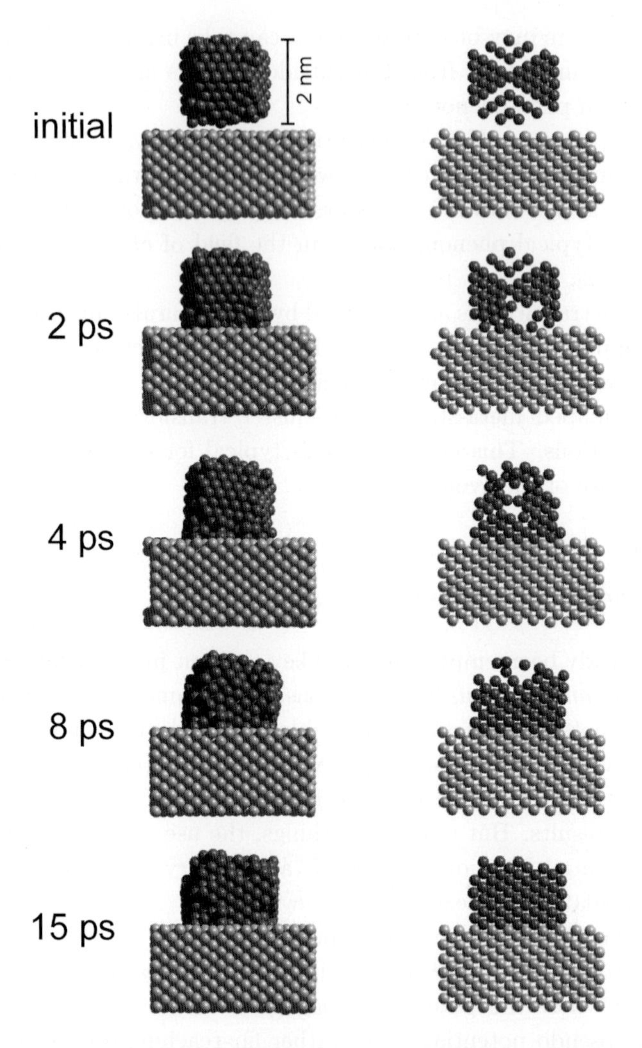

Fig. 5.28 Cluster deposition on the (001) surface of a thin aluminium film at 300 K (the left row shows the situation from the front, the right row illustrates the middle layer): Initially a stable Al_{500} cluster (the same as shown in Figs. 5.12 and 5.13) is put within reach of a (001) structured aluminium surface. After 2 ps the cluster has established full contact with the surface. The initially poly-crystalline structure which can be clearly recognized from the atomic arrangement of the middle layer starts to recrystallize from the bottom. This process continues with time and after 15 ps the former dodecahedral shaped poly-crystalline Al_{500} cluster has the form of a cuboid. The front, back, left and right faces are (111), the bottom and top surfaces are (001) structured. This is due to a lattice contraction along the vertically axis. The final cluster structure meets those of a perfect single crystal with several vacancies and ad-atoms at the surface. The poly-crystalline structure has completely vanished.

nanostructured materials work at rather high temperatures [FZJ, 1998] and, in addition, cause a lot of cluster collisions, formations of meta-stable and pseudo stable clusters are not expected to be common products of today's materials synthesis. Therefore, at first glance, the considered MD studies of the *bifurcation phenomenon* for the example of single isolated nano-clusters seems to be of academic interest only. But in Figs. 4.1 and 4.2, it has already been demonstrated that nano-clusters which interact with surfaces show structure transformations too.

An additional example is given in Fig. 5.28. It demonstrates the recrystallization process of a multi-crystalline Al_{500} cluster in the stable state (it is the same as shown in Figs. 5.12 and 5.13) due to the interaction with an Al (001) surface at 300 K.

Furthermore, the synthesis of nanostructured materials (e.g. by cluster compression and sintering) is dominated by cluster–cluster interactions. In this connection, recrystallization effects and stability considerations play a major role. The MD study illustrated in Fig. 5.29 gives some impressions to this topic.

In this way molecular dynamic investigations of basic cluster characteristics — like the considered *bifurcation phenomenon* — in combination with more complex material models might give fresh stimulus to material science.

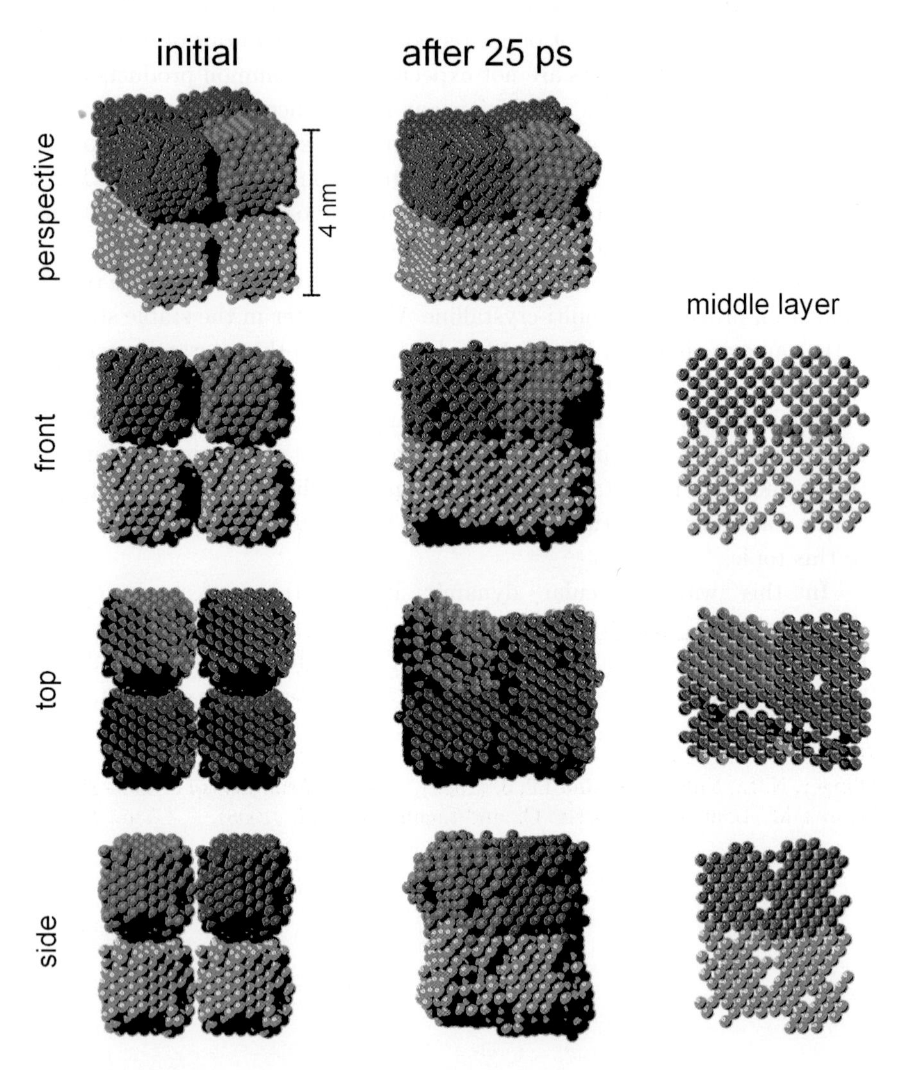

Fig. 5.29 Eight free Al_{500} clusters symmetrically arranged merge to a single Al_{4000} structure at 50 K (the clusters are colored different for a better relation of the particles): This is a simple model to simulate the process that takes place during the synthesis of nanostructured materials. As can be seen, the poly-crystalline structures vanish due to a re-crystallization process. But this does not lead to a perfect structure. Though the outer shape is similar to the Al_{2048} clusters of Fig. 5.21, there are significant irregularities at the surface as well as in the inner lattice structure. The pictures of the middle layers clearly show vacancies and dislocations.

Bibliography

Ahlrichs, R., and v. Arnim, M. (1995), "TURBOMOLE, Parallel Implementation of SCF, Density Functional, and Chemical Shift Modules", in: "Methods and Techniques in Computational Chemistry", *METECC-95*, E. Clementi, G. Corongiu (eds.);
also in: http://www.chemie.uni-karlsruhe.de/PC/TheoChem.

Ahlrichs, R., and Elliott, S. D. (1999), *PCCP* **1** (1), 13.

Alig, I., Kröhn, M., and Hentschke, R. (2000), "Modellieren von Molekülen", in: *Spektrum der Wissenschaft — Digest: Moderne Chemie II.*

Allen, R. E., and F. W. de Wette (1969), *J. Chem. Phys.* **51**, 4820.

Allen, M. P., Tildesley, D. J. (1990), "Computer Simulation of Liquids", Oxford Science Publications.

Allinger, N. L. (1977), *J. Am. Chem. Soc.* **99**, 8127–8134.

Allinger, N. L., Yuh, Y. H., and Lii, J. (1989), *J. Am. Chem Soc.* **111**, 8551–8582.

Alvarez, M., Lomba, E., Martín, C., and Lombardero, M. (1995), *J. Chem. Phys.* **103**, 3680.

Andersen, H. C. (1980), *J. Chem. Phys.* **72**, 2384.

Andersen, H. C., Allen, M. P., Bellemans, A., Board, J., Clarke, J. H. R., Ferrario, M., Haile, J. M., Nóse, S., Opheusden, J. V., and Ryckaert, J. P. (1984), *Rapport d'activité scientifique du CECAM*, 82–115.

Animalu, A. O. E., and Heine, V. (1965), *Philos. Mag.* **12**, 1249.

Ashcroft, N. W. (1966), *Phys. Lett. 23*, 48.

Autodesk Inc. (1992), Autodesk Animation (AA) Player for Windows.

Autodesk Inc. (1998), Genius Desktop,
Mechanical Desktop 3,
AutoLISP for AutoCAD R14, http://www.autodesk.com,
(all product names are registered trademarks).

Balamane, H., Halicioglu, T., and Tiller, W. A. (1992), *Phys. Rev. B* **46**, 4.

Barker, J. A., Watts, R. O., Lee, J. K., Schafer, T. P, and Lee, Y. T. (1974), *J. Chem. Phys.* **61**, 8.

Beazley, D. M., and Lomdahl, P. S. (1994), *Par. Comput.* **20**, 173.

Beeler (Jr.), J. R., and Kulcinski, G. L. (1972), in [Gehlen *et. al*, 1972].

Beeman, D. (1976), *J. Comput. Phys.* **20**, 130.
Beloussov, L. V. (2002), in: "What is Live?", H.-P. Dürr, F.-A. Popp, W. Schommers (eds.), Series on the Foundations of Natural Science and Technology — Vol. 4, World Scientific;
also in: (2000) "Elemente des Lebens", H.-P. Dürr, F.-A. Popp, W. Schommers (eds.), Die Graue Edition, SFG, Kusterdingen, Germany.
Berendsen, H. J. C., and van Gunsteren, W. F. (1986), in [Ciccotti *et. al*, 1986].
Boisvert, G., and Lewis, L. J. (1997), *Phys. Rev. B* **52**, 7643.
Born, M., and Oppenheimer, J. R. (1927), *Ann. Phys.* **84**, 457.
Box, G. E. P., and Muller, M. E. (1958), *Ann. Math. Stat.* **29**, 610.
Buckingham, R. A. (1938), *Proc. Roy. Soc. A* **168**, 264.
Buckingham, R. A. (1961), *J. Plan. Space Sci.* **3**, 205.
Burkert, U., and Allinger, N. L. (1982), "Molecular Mechanics", *ACS Monogr.* 177, American Chemical Society, Washington.
Casanova, *et al.* (1970), *Molecul. Phys.* **18**, 589.
Ciccotti, G., and Hoover, W. G. (ed.) (1986), "Molecular-dynamics simulation of statistical-mechanical systems", *International School of Physics "Enrico Fermi" (1985: Varenna, Italy)*, North-Holland Physics Publishing, Amsterdam, The Netherlands.
Clark, T. (1985), "A Handbook of Computational Chemistry", John Wiley & Sons, New York.
Croxton, C. A. (1980), "Statistical Mechanics of the Liquid Surface", Wiley, Chichester.
Dahlquist, G., and Björck, A. (1974), "Numerical Methods", Englewood Cliffs, New Jersey.
Daw, M. S., and Baskes, M. I. (1983), *Phys. Rev. Let.* **50**, 1285.
Dawid, A., and Gburski, Z. (1998), *Phys. Rev. A* **58**, 1.
della Valle, R. G., and Venuti, E. (1998), *Phys. Rev. B* **58**, 1.
Drexler, K. E. (1981), "Molecular Engineering: An approach to the development of general capabilities for molecular manipulation", *Proc. Nat. Acad. Sci. USA* **78**, 5275.
Drexler, K. E., Peterson, C., Pergamit, G. (1991), "Unbounding the Future: The Nanotechnology Revolution", William Morrow and Company, Inc., New York.
Drexler, K. E. (1992), "Nanosystems: Molecular Machinery, Manufacturing, and Computation", John Wiley.
Eberhardt, W., in: [FZJ, 1998], 9.
Egelstaff, P. A. (1967), "An Introduction to the Liquid State", Academic Press, London.
Farquhar, I. E. (1964), "Ergodic Theory in Statistical Mechanics", Interscience Publishers, New York.
Feynman, R. P. (1960), *Eng. Sci.* **23**.
Fincham, D., and Ralston, B. J. (1981), *Comput. Phys. Commun.* **23**, 127.
Fincham, D., and Heyes, D. M. (1982), *CCP*5 *Quart.* **6**, 4.
Fock, V. (1930), *Z. Phys.* **61**, 126.

Foiles, S. M., Baskes, M. I., and Daw, M. S. (1986), *Phys. Rev. B* **33**, 7983.

Friedman, H. L. (1985), "A Course in Statistical Mechanics", Prentice-Hall, Englewood Cliffs, New Jersey.

FZJ (1998), "Physik der Nanostrukturen", *Schriften des Forschungszentrums Jülich: Reihe Materie und Material*, Band 1, Jülich, Germany.

Gear, C. W. (1966), *Rep. ANL*-7126, Argonne National Laboratory.

Gear, C. W. (1971), "Numerical Initial Value Problems in Ordinary Differential Equations", New Jersey.

Gehlen, P. C., Beeler (Jr.), J. R., Jaffee, R. I. (eds.) (1972), "Interatomic Potentials and Simulation of Lattice Defects", Plenum Press, New York.

Geldart, W., and Vosko, S. H. (1966), *Canad. J. Phys.* **44**, 2137.

Girifalco, L. A., and Weizer, V. G. (1959), *Phys. Rev. 114*, 687.

Greiner, W. (1993), "Quantentheorie: Spezielle Kapitel", *Theoretische Physik*, Bd. 4A, Harri Deutsch, Thun, Frankfurt am Main, Germany.

Greiner, W., Neise, L., Stöcker, H. (1993), "Thermodynamik und Statistische Mechanik", *Theoretische Physik Bd.* 9, Harri Deutsch, Thun, Frankfurt am Main.

Halicioglu, T., and Pound, G. M. (1975), *Phys. Stat. Solidi A* **30**, 619.

Hamming, R. W. (1973), "Numerical Methods for Scientists and Engineers", McGraw-Hill, New York.

Han, J., Globus, A., Jaffe, R., and Deardorff, G. (1997), *Nanotechnology* **8**, 95.

Harrison, W. A. (1966), "Pseudopotentials in the Theory of Metals", W. A. Benjamin, Inc., New York, Amsterdam.

Hartree, D. R. (1955), "The Calculation of Atomic Structures", Wiley, New York.

Haug, E. J. (1991), "Real-time Integration Methods for Mechanical System Simulation", Springer.

Heine, V., and Abarenkov, I. V. (1964), *Philos. Mag.* **9**, 451.

Heyes, D. M., and Singer, K. (1982), *CCP*5 *Quart.* **6**, 11.

Hirschfelder, J. O., Curtis, C. F., and Bird, R. B. (1954), "Molecular Theory of Gases and Liquids", Wiley, New York.

Hockney, R. W. (1970), *Meth. Comput. Phys.* **9**, 136.

Hofer, E., Lunderstädt, R. (1975), "Numerische Methoden der Optimierung", R. Oldenbourg, München, Wien.

Hubbard, J. (1958), *Proc. Roy. Soc.* A **243**, 336.

Jena, P., Rao, B. K., and Khanna, S. N. (eds.) (1987), "Physics and Chemistry of Small Clusters", *NATO ASI Ser. B: Phys.* Vol. 158, Plenum Press, New York.

Kirkpatrick, S., Gelatt, C. D., and Vecci, M. P. (1983), *Science* **220**, 671.

Kohn, W., Sham, L. J. (1965), *Phys. Rev. 140*, A1133.

Komvopoulos, K., and Yan, W. (1997), *J. Appl. Phys.* **82**, 10.

Koonin, S. E., Meredith, D. C. (1990), "Computational Physics", Addison Wesley.

Landau, L. D., and Lifshitz, E. M. (1959), "Quantum Mechanics", *Course of Theoretical Physics*, Vol. 3, Pergamon Press, London, Paris.

Lapidus, L., and Seinfeld, J. H. (1971), "Numerical Solution of Ordinary Differential Equations", New York.

Lomdahl, P. S., Tamayo, P., Grønbech–Jensen, N., and Beazley, D. M. (1993), *Proc. Supercomput. '93*, IEEE Computer Society Press, Silver Spring, MD, 520.

Lucas, K. (1991), "Applied Statistical Thermodynamics", Springer.

Ludwig, W. (1969), "Diploma Work", University of Münster, Münster, Germany.

Maradudin, A. A., Montroll, E. W., Weiss, G. H., and Ipatova, I. P. (1967), "Theory of Lattice Dynamics in the Harmonic Approximation", *Solid State Phys.*, Academic Press, New York, Suppl. 3.

March, N. H. (1968), in: "Theory of Condensed Matter", International Atomic Agency, Vienna.

Marsaglia, G. (1972), *Ann. Math. Stat.* **43**, 645.

Mason, D. K. (1993), DTA (Dave's TGA Animation Program) Rel. 2.0.7, P.O. Box 181015, Boston, Massachusetts 02118, USA, ftp://ftp.povray.org/pub/povray/utilities/dta/

Mason, D. K., and Enzmann, A. (1993), "Making Movies on Your PC", The Waite Group, 1993.

Merkle, R. C. (2001), "Nanotechnology: What will it mean?", *IEEE Spectrum* 1, 19;
(2001), "That's impossible! How good scientists reach bad conclusions", http://www.zyvex.com.

Messiah, A. (1990), "Quantenmechanik", Vol. 2, Walter de Gruyter, Berlin, New York.

Milanski, J. (2000), "Design Takes Charge — Designing in the Wake of Nanotechnology", Illinois Institute of Technology, http://iit.id.edu/ milanski.

Morse, P. M. (1929), *Phys. Rev.* **34**, 57.

The National Nanotechnology Initiative (NNI), http://nano.gov.

Nordsieck, A. (1962), *Math. Comput.* **16**, 22.

Physikalisch Technische Bundesanstalt (1985), "Die gesetzlichen Einheiten in Deutschland", Braunschweig und Berlin;
also in: http://www.ptb.de.

Policy Research Project on Anticipating the Effects of New Technologies (1989), *Assessing Molecular and Atomic Scale Technologies* (*MAST*), Austin, University of Texas.

Powles, J. G., Evans, W. A. B., and Quirke, N. (1982), *Mol. Phys.* **46**, 1347.

Rahman, A. (1964), *Phys. Rev.* **136**, A405.

Rahman, A., and Stillinger, F. H. (1971), *J. Chem. Phys.* **55**, 3336.

Ralston, A., and Wilf, H. S. (1967), "Mathematische Methoden für Digitalrechner", R. Oldenbourg, München.

Regis, E. (1995),"Nano!", Little Brown.

Rey, C., Gallego, L. J., García-Rodeja, J., Alonso, J. A., and Iñiguez, M. P. (1993), *Phys. Rev. B* **48**, 8253.

Rice, S. A., and Gray, P. (1965), "The Statistical Mechanics of Simple Liquids", *Monographs in Statistical Physics and Thermodynamic*, I. Prigogine (ed.), Interscience Publishers.

Rieth, M., Schommers, W., Baskoutas, S., and Politis, C. (1999), *NACHRICHTEN — Forschungszentrum Karlsruhe* **31**, 2–3.

Rieth, M., Schommers, W., and Baskoutas, S. (2000), *Mod. Phys. Lett. B* **14**, 621.

Rieth, M., Schommers, W., Baskoutas, S., Politis, C., and Jannussis, A. (2001), *Chin. Phys.* **10**, 137;
Rieth, M., Schommers, W., Baskoutas, S., and Politis, C. (2001), *Chin. Phys.* **10**, 132.

Rieth, M., Schommers, W. (2002), in: "What is Live?", H.-P. Dürr, F.-A. Popp, W. Schommers (eds.), *Series on the Foundations of Natural Science and Technology*, Vol. 4, World Scientific.

Roco, M. C., and Sims, W. (ed.) (2001),"Societal Implications of Nanoscience and Nanotechnology", NSET Workshop Report, Bainbridge National Science Foundation, Arlington, Virginia.

Salacuse, J. J., Schommers, W., and Egelstaff, P. A. (1986), *Phys. Rev. A* **34**, 1516.

Sauer, J. (2000), "Chemie aus dem Computer", in: *Spektrum der Wissenschaft — Digest: Moderne Chemie II.*

Schommers, W. (1976), *Z. Phys. B* **24**, 171–175.

Schommers, W. (1977), *Phys. Rev. Lett.* **38**, 1536;
(1980), *Phys. Rev. B* **21**, 847;
(1980), *Phys. Rev. B* **22**, 1058.

Schommers, W. (1986), in: "Structures and Dynamics of Surfaces I", *Topics in Current Physics*, Vol. 41, W. Schommers and P. von Blanckenhagen (eds.), Springer-Verlag, Berlin, Heidelberg.

Schommers, W. (1987), in: "Structures and Dynamics of Surfaces II", *Topics in Current Physics*, Vol. 43, W. Schommers and P. von Blanckenhagen (eds.), Springer-Verlag,

Schommers, W., Mayer, C., Göbel, H., and von Blanckenhagen, P. (1995), *J. Vac. Sci. Technol. A* **13**, 3.

Schommers, W., and Rieth, M. (1997), *J. Vac. Sci. Technol. B* **15**, 1610.

Seifert, G. (1998), in [FZJ, 1998], C4.1.

Sham, L. J. (1965), *Proc. Roy. Soc. A* **283**, 33.

Shaw, R. W., and Harrison (1967), W. A. (1967), *Phys. Rev.* **163**, 604.

Shaw, R. (1968), *Phys. Rev.* **174**, 769.

Shaw, R. W., and Pynn, R. (1969), *J. Phys. C* **2**, 2071.

Siegel, R. W. (1997), *Spektrum der Wissenschaft* **3**, 62.

Singwi, K. S., Tosi, M. P., Land, R. H., and Sjolander, A. (1970), *Phys. Rev. B* **1**, 1044.

SOFTreat (2001), 3D Stereo Image Factory PLUS V2.5, 1450 Long Mill Road, Youngsville, North Carolina 27596.

Stoddard, S. D., and Ford, J. (1973), *Phys. Rev. A* **8**, 1504.

Stott, M. J., and Zaremba, E. (1980), *Phys. Rev. B* **22**, 1564.

Streett, W. B., Tildesley, D. J., and Saville, G. (1978), in: "Computer Modelling of Matter", P. Lykos (ed.), *ACS Symp. Ser.*, Vol. 86, American Chemical Society, Washington.

Swope, W. C., Andersen, H. C., Berens, P. H., and Wilson, K. R. (1982), *J. Chem. Phys.* **76**, 637.

Thompson, S. M. (1983), *CCP5 Quart.* **8**, 20.

Torrens, I. M. (1972), "Interatomic Potentials", Academic Press, New York, London.

van Gunsteren, W. F., and Berendsen, H. J. C. (1977), *Mol. Phys.* **34**, 1311.

Verlet, L. (1967), *Phys. Rev.* **159**, 98.

von Blanckenhagen, P., and Schommers, W. (1987), in: "Structures and Dynamics of Surfaces II", *Topics in Current Physics*, Vol. 43, W. Schommers and P. von Blanckenhagen (eds.), Springer-Verlag, Berlin, Heidelberg.

von Neumann, J. (1951), *US Natl. Bur. Stand. Appl. Math.* **12**, 36.

Voter, A. F., and Chen, S. P. (1987), in: "Characterization of Defects in Materials", R. W. Siegal, J. R. Weertman, and R. Sinclair (eds.), *MRS Symposia Proc.* No. 82, Materials Research Society, Pittsburgh.

Wirdelius, H. (2000), SKI Report 00:29, Swedish Nuclear Power Inspectorate, Mölndal, Sweden.

Young, C., and Wells, D. (1994), "Ray Tracing Creations", Waite Group Press.

Young, C. (1997), Persistence of Vision Ray Tracer (POV–Ray) Version 3.1, 3119 Cossell Drive, Indianapolis, IN 46224, USA,
http://www.povray.org
(all product names are registered trademarks).

Index